# EATS WITH OATS

## The NEW Soluble Fibre Cookbook

# EATS WITH OATS

## The NEW Soluble Fibre Cookbook

Janette Marshall

**W. Foulsham & Co. Ltd.**
London · New York · Toronto · Cape Town · Sydney

W. Foulsham & Company Limited
Yeovil Road, Slough, Berkshire, SL1 4JH

ISBN 0–572–01352–3



Printed in Spain by Cayfosa. Barcelona
Dep. Leg. B-5413-1986

# Contents

| | | Page |
|---|---|---|
| | *Introduction* | 6 |
| 1. | What is fibre? | 7 |
| 2. | How to add more fibre to your diet | 12 |
| 3. | Where do oats come from? | 15 |
| 4. | Balancing the diet | 17 |
| 5. | Oats down on the health farm | 21 |
| | *Recipes* | |
| 6. | Breakfast | 27 |
| 7. | Main meals | 41 |
| 8. | Desserts | 65 |
| 9. | Bread and Teatime Treats | 84 |
| 10. | Snacks | 96 |
| 11. | Biscuits | 109 |
| 12. | Christmas and Easter | 115 |
| | *Index* | 127 |

# INTRODUCTION

Oats have been around for a long time, but in recent years with a switch away from porridge to bran-based breakfast cereals the amount we have eaten has declined. But now oats are right back in dietary fashion.

Why? Because oats are high in fibre and what is more they contain a very special type of fibre that scientific research has shown helps control blood pressure, protect against heart disease and is especially good for diabetics.

Until now the interest in fibre has centred on bran. We all know the advantages of bran in keeping a healthy digestion and avoiding constipation and the more serious problems that can develop from it. But what we haven't known before is that there are two types of dietary fibre: soluble and insoluble.

Bran is insoluble and although valuable in its own way it should be eaten with foods which contain soluble fibre. Which is where oats come in because they (and beans) are the best sources of soluble fibre, making them an invaluable part of the diet for everyone who wants to keep fit and healthy.

And what's more oats are amazingly versatile. So, even if you don't like porridge or muesli (and with the recipes in this book you are sure to find one of these to your taste) there are plenty of other ways during the day when you can make sure of getting your oats!

# *Chapter 1*
# WHAT IS FIBRE?

Fibre is the substance that makes up the cell walls of plants. If you imagine a cell to be like a cardboard box the fibre is the equivalent of the sides of the box. Pile several on top of each other and you have a simplified illustration of the way the fibre in plant walls enables the plant to stay standing up. Animal cells do not contain fibre, but that does not mean that they are always falling over! Humans and animals are kept upright by their skeleton. Muscles are attached to the bones to move the body about. This basic difference is important because it highlights the fact that fibre is only available in plant foods. Meat, fish, dairy produce – although valuable in the diet for other reasons – contains no fibre.

Not all plants contain the same amount or type of fibre. Trees, for example, have 'high fibre' trunks and branches, but they are of no dietary use to man. The types of plants that are high in fibre and useful in the diet often give themselves away by their texture. Pulses (beans, peas and lentils) are good sources of fibre and so are cereals (wheat, oats, barley, rice, rye, etc.) and dried fruits (dates, prunes, apricots, figs) are too. Fresh fruit and vegetables are also good sources of fibre.

### Soluble and Insoluble Fibre

To illustrate the different roles of soluble and insoluble fibre let's look at a simple experiment. Put some bran in a cup of water and stir it. After a short while the bran swells as it absorbs the water making a soft, bulky substance. This is insoluble fibre at work. Its ability to soak up water is very good for our digestion because it gives bulk to the food we eat which helps us in several ways.

Firstly, it helps us maintain a healthy digestion because by making the food soft and bulky the muscles are able to push it through the body easily and quickly to dispose of waste products. This helps prevent many common health problems associated with constipation. Secondly, the bulking action of fibre makes us feel fuller after we have eaten, so helping us to eat less (fewer calories) and stay slim. As long as high fibre foods are low in fats and sugars they can give us a lot of bulk for few calories, and unrefined high fibre foods also have the vitamins and minerals in their natural state unlike processed and refined foods.

To illustrate the slimming effect consider that it takes 15 minutes to eat three apples and the same number of calories can be obtained from apple juice in $1\frac{1}{2}$ minutes. These are the best known roles of fibre in the diet, but it is the more recent discoveries that have shown us the importance of the second type of fibre – soluble fibre which is found in oats, beans and other pulses.

Instead of helping directly in the efficient working of the digestive system – like bran does – the gummy soluble fibre found in oats and also in beans helps keep us healthy in a different way. It can in fact help lower the levels of harmful cholesterol in the bloodstream and also help regulate blood sugar levels which is especially helpful in prevention and dietary maintenance of diabetes. It may also help prevent gallstones and high blood pressure because of its effect on blood sugar and fat levels.

Today most of us are aware of the risks of heart disease. Along with taking proper exercise, not smoking and watching our diet in general we have been advised by doctors to be especially careful that we cut down on the amount of fats and especially saturated fats. This means the cholesterol containing animal fats found in meat, milk, butter and cheese. The way in which oats can help regulate these fats is through the soluble fibre they contain which has the effect of lowering the amount of cholesterol in our blood and regulating the amount of sugar in the blood – two elements that are thought to lead to the silting up of our arteries by arterial plaque. The plaque is deposited on the sides of the arteries narrowing the channel through which the blood is pumped. This increases the chances of thrombosis in which blood clots get stuck and stops oxygenated blood reaching the heart.

**Want to know more?**

If you would like to know more we can now look at the technical side of how soluble fibre works.

First let's introduce Dr James Anderson, professor of medicine and clinical nutrition at the University of Kentucky, Lexington, USA. It is his work that has shown how oat fibre can help prevent and control diabetes. Using his HCF Diet which is high in fibre and high in carbohydrates (starchy foods) he was able to reduce insulin dependency by half in a group of diabetic patients he treated with this diet. They were all maturity-onset diabetics, which meant they developed the disease later in life, often as a result of poor diet.

Dr Anderson found his diet lowered the insulin requirements of the patients and decreased the swings in blood sugar levels experienced by them. 'But more importantly,' he says 'our patients find the diets practical and palatable . . . they actually *like* the diets.'

The reason that oats and beans are able to stop swings in blood sugar levels is that they are unrefined and therefore contain soluble fibre. This slows down the digestion of the carbohydrates and provides a steadier stream of energy to the body in the form of a steadier stream of sugars into the bloodstream. This avoids the sudden demands on the pancreas to produce insulin which a diet high in refined carbohydrates and sugary foods does. This type of diet is not only good for diabetics or pre-diabetics, but it is good for all of us because it does not put our bodies under strain and because the natural and unrefined foods are naturally high in fibre and vitamins and minerals.

It was while working with diabetic patients that Dr Anderson's team also noticed that not only were their patients' blood sugar levels smoothed out but also the levels of fat in their blood began to drop. In one of his trials Dr Anderson gave eight men with high levels of cholesterol in their blood 100g (4 oz) of oatbran each day. The men also ate the same diet for a certain time without the addition of oatbran. During the time they had the oatbran supplement the levels of harmful cholesterol were lowered by 13 per cent, and that was without any other dietary modification such as lowering the amount of fat in the diet. We say 'harmful' cholesterol because there are two main types of cholesterol: low density lipoprotein (LDL) and high density lipoprotein (HDL). It is the LDL which forms arterial plaque leading to clogged arteries and reduced flow of blood to and from the heart, leading to increased risk of heart attacks. It is exactly this harmful substance that the soluble fibre in the oatbran, in the form of a polysaccharide called beta-glucan, combines with to remove from the body. The HDL, which is thought to be beneficial to the body by helping prevent arterial plaque depostis, is unaffected by the special properties of oatbran.

Soluble fibre seems especially good at reducing the levels of harmful cholesterol and there are two ways in which it does it. Firstly, soluble fibre binds with bile acids and removes them from the body. Bile acids are made from cholesterol in the liver and are needed for the body to digest fat. So, by binding with them and removing them, less cholesterol gets into the bloodstream because it is not being digested, due to the reduced amount of bile acids around to do the job. Secondly, most of the cholesterol in the body is made by the liver. Soluble fibre is used by bacteria in the gut to produce substances that are taken to the liver where they actually switch off cholesterol production. Both these help lower blood fat levels.

Dr Anderson claims that by eating soluble fibre in the form of oats and beans cholesterol levels can be lowered by as much as 30 per cent, which means less need to resort to drugs when controlling high cholesterol problems and, most importantly for most of us, it may also mean that oats and beans can prevent such problems arising in the first place. But like any other food oats are not a 'wonder drug' or 'wonder food' working alone. To be effective they have to be part of a sensible eating and exercise regime.

Adding oats to the diet can also help prevent the common problem of high blood pressure. Studies of diets have shown people on a high fibre diet, especially vegetarians, are less likely to suffer from high blood pressure and Dr Anderson's work has confirmed this. Dietary management of high blood pressure also means cutting down on salt and both these methods are useful because they reduce the need for drugs with their undesirable side-effects.

Soluble fibre may also reduce the risk of gallstones because, as we have seen, it reduces the amount of bile acids in the system and bile acids and cholesterol combine to form gallstones.

**To sum up**

So, for a diet to keep us healthy and to help those with specific health problems, we need to make sure we get enough of both types of fibre – the soluble and the insoluble – because they both do valuable and different jobs in keeping us fit and healthy. We need the insoluble bran-type fibre for preventing constipation and other digestive and related problems and we need soluble fibre to help maintain a healthy balance of fats and sugars in the blood.

We should make sure that our diet is made up of mainly unrefined carbohydrates which are high fibre, starchy foods like potatoes, beans, peas, corn, wholemeal breads and pasta, cereals like oats and brown rice and other grains. They are not fattening and bad for us – that is now a myth consigned to gather dust by today's nutritionists. These are the foods that give us energy, fibre and vitamins and minerals. We also need protein from lean meat or fish, vegetarian dishes, some cheese, eggs and skimmed milk. And we need a small amount of vegetable fats or marine oils from oily fish to provide us with the essential fatty acids that our body cannot make.

And with the knowledge of these new soluble fibre discoveries we need to take a good look at oats, which have long been neglected in favour of wheat and bran-based cereals, not only for the contribution they can make to the breakfast table in porridge, muesli, muffins and scones but also the many other ways in which we can use them throughout the day to add to a balanced diet with a rich variety of flavours and textures.

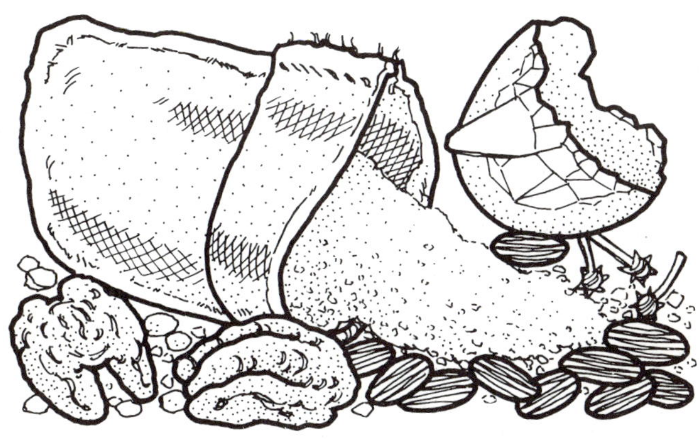

*Chapter 2*

# HOW TO ADD MORE FIBRE TO YOUR DIET

Sprinkling bran on everything you eat or relying on a bran breakfast cereal as many people do is not the best way to increase the amount of fibre in your diet. Adding bran to a diet based on refined foods such as white bread and flour products like cakes and pastries, or meat and dairy produce will not help you avoid all the health problems associated with a low fibre diet.

The best way to eat a high fibre diet is to rely more on natural foods that are as unprocessed as possible, and therefore have not had all their fibre removed by food processing, which happens with white flour or sugar. It also means watching the amount of low fibre and high fat foods you eat, such as meats, eggs, cheese, confectionery, sweet sauces, jams, jellies, and sugary foods.

The sort of natural foods that include naturally present fibre are:

Wholemeal bread, crispbreads, wholemeal crackers, pitta bread.

Oats, such as porridge, muesli, breakfast dishes, oatcakes, biscuits, oat pastries, etc.

Wholemeal pastry with oats and oatmeal added for a change.

Brown rice and other whole grains like millet, barley, whole wheat, bulghur.

Wholemeal and buckwheat pasta.

Pulses such as dried beans, peas, lentils, chick peas – even baked beans!

Fresh fruit, washed but unpeeled.

Fresh vegetables, washed but unpeeled where possible. Eat the skins of potatoes (or sweet potatoes) baked in their jackets, and choose the fibrous varieties of lettuce and greens rather than the floppy ones.

Dried fruits such as prunes, dates, figs, apricots, peaches, raisin, sultanas, currants, coconut.
Nuts and seeds – in moderation because they are high in fat and therefore high in calories.

**How much fibre to aim for**

It is difficult to state an exact amount of fibre to be eaten each day. No one wants to bother to add up their fibre intake in the same way that some slimmers do their calorie intake, because that can become obsessive and it takes the enjoyment out of eating. So it makes sense to revise your pattern of eating to include many of the foods listed for adding fibre to your diet and to cut down on the fatty, low fibre foods. The fibre content of your diet will then take care of itself.

There is no 'officially recognised' amount of fibre set down in government recommended daily intakes as there is for some nutrients like vitamins and minerals. However, Dr Denis Burkitt and Captain Surgeon Cleave, pioneer fibre researchers,have found that as little as 15–25g (1/2–1 oz) of fibre is eaten daily in the typical Western diet as compared with 40–60g (1/2–2 oz) daily in undeveloped countries where the diseases linked to low fibre diets do not exist. Generally accepted figures seem to point to about 30–35g ($1\frac{1}{2}$ oz) a day and it is not recommended to exceed 50g (2 oz) a day.

Dr Anderson has recommended that patients who leave hospital after treatment for diabetes or heart disease eat one bowlful of porridge a day and three oat muffins to control their cholesterol and blood sugar levels. In his first trials with oats Dr Anderson ate 75g (3 oz) of oatbran a day in the form of breakfast cereal and oat muffins. In five weeks his cholesterol level dropped by 40 per cent.

'After this,' he says, 'I began on *real* people!' He continues, 'Our next target was people who had twice the serum cholesterol level of healthy people. We wondered if oats would work for them if they still ate eggs, meat and the ingredients of a typical Western diet'. By taking 75g (3oz) of oatbran daily they reduced cholesterol levels by 20 per cent, between ten days and three weeks, thereby reducing heart disease risks.'

'After I found I could lower my cholesterol level I became a "believer" and for six years (when I am at home) I have eaten an oat muffin and a bowlful of porridge made with skimmed milk each morning, and for lunch I have two oat muffins to keep a healthy cholesterol level.'

'I like this diet and what's more my patients like it, so when they escape from hospital they can follow a diet that emphasises porridge, oatbread, oat scones or muffins and sustain their improvements. Two years after they have left hospital their serum cholesterol level remains the same, or lower, than when they left hospital.

'There is a high risk of heart disease among men at certain vulnerable ages. If they begin to eat porridge it can act as a preventive and lower their risk of heart disease by between 10 and 15 per cent.'

*Chapter 3*

# WHERE DO OATS COME FROM?

Oats have traditionally been associated with Scotland because they are a hardy crop that survives the cold weather of the north. In the seventeenth and eighteenth centuries they gradually replaced barley and rye because they grew better, but the improved agricultural methods of the nineteenth century saw wheat replacing oats.

Today oats are grown in almost every part of the world but the most important areas of production are North America, Scandinavia, Russia, Great Britain and Australia. Oat grains have been found in archaeological remains in all parts of England and on Germanic and old Nordic sites. Since the thirteenth century they have been popular in Britain, but today there are fewer oats than ever being grown in Britain. In fact the British crop has declined between 65 and 75 per cent in the last 15 years.

## Producing Porridge Oats

To make porridge oats the grain is first put through a process called roasting or kilning. The oat passes through a machine which blows hot gases from a furnace over it. This reduces the moisture content of the grain from 16 to 6 per cent. The grain is then cooled and dehulled by a machine which hurls the grains at a rubber wall to spring the kernel or oat groat from the husk or shell.

At the same time some of the second skin (or bran) beneath the husk comes away and with it a little germ. However, there is still some bran and other fibre in the remaining groat. The groats, husks and second skin pass over an aspirator which blows away the husk and second skin and leaves the groat to continue its journey to become porridge oats or oatmeal. The oat bran and oat germ is separated from the husk and cleaned, but nothing is added or taken away. Previously oat bran and oat germ would have been used as animal feed, just as wheat bran used to be before its dietary significance was realised. Now it can be bought just like wheat bran.

Meanwhile the groats are cut into three pieces to produce a coarse pinhead meal before being stabilised. The process of stabilisation is a closely guarded secret among oat millers, but they say nothing is added or taken away to the groat during the process. It involves steam treating the groats to inactivate the lipase enzyme which would otherwise cause degeneration of the fats and shorten the products' shelf life. Stabilisation is a finely balanced process because over-heating will cause off flavours.

Before the breakthrough of stabilisation, oatmeal and oats could be stored for only two months in an airtight tin. They also took longer to cook. The steam treatment process of stabilisation has reduced cooking time from around 45 minutes to about five or ten, depending on personal preference. After being stabilised the pinhead meal is either cooled and granulated or flaked to make oatmeal. It may also be rolled flat into porridge oats. Jumbo oats or oatflakes are whole oats rolled flat into porridge oats.

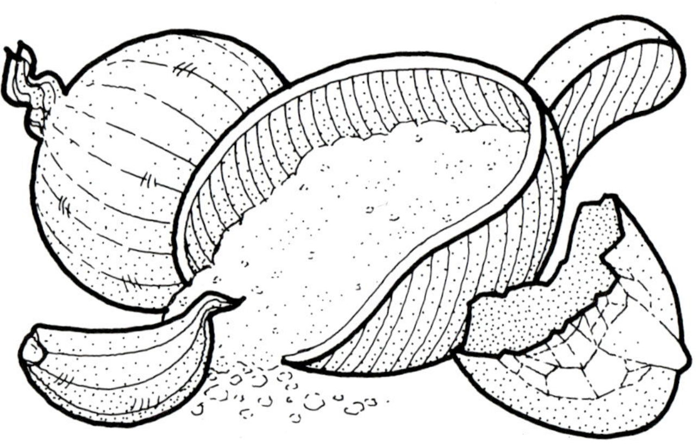

# *Chapter 4*
# BALANCING THE DIET

Man cannot live on fibre alone! We need carbohydrates, protein and a little fat, vitamins and minerals in order to be healthy and fit, but there are five basic principles to healthy eating:

***1. Cut down on fat***
Aim for around less than 75g (3 oz) a day including the invisible fats in cakes and pastries, and cooking oils. Make the most of that fat intake high in polyunsaturates.

***2. Cut down on sugar***
We don't *need* any 'table' sugar such as that added to tea, coffee or sprinkled on breakfast cereals and desserts. Try to avoid it in processed foods by reading the labels and remembering that glucose, syrups, dextrose, etc. are all 'sugars'. We get all the energy we need from unrefined carbohydrate foods.

***3. Cut down on salt***
Make a start by not sprinkling it on food at the table and cutting down the amount you use in cooking – it is possible to wean yourself off salt and rediscover the *real* taste of food. Watch out for hidden salt in processed foods by reading the labels.

***4. Cut down on meat***
It's not necessary for health, but if you use meat choose the leaner cuts and white meats and game which are lower in fat. Don't add fat when cooking; roasting is very successful without extra fat. Fish is a low-fat source of protein, which can also be obtained by combining two of the three groups of vegetable protein foods: nuts and seeds, cereals and pulses.

***5. Eat more fibre, vitamins and minerals***
This means going for more fresh fruit and vegetables, choosing wholemeal bread, brown rice and pasta, beans and pulses and good whole natural foods like oats!

It's also worth mentioning that although diet is very important and can mean the difference between feeling 'all right' and 'great!' there are other things to consider. You can eat the best diet in the world but if you smoke you are risking your health. So, try to cut down and aim to stop smoking. Exercise too is important. The minimum daily dose set by experts is 15 minutes of 'aerobic' exercise. This means exercise which raises the pulse and therefore the metabolic rate. This type of exercise continues burning off the calories long after you have finished exercising because the metabolic rate continues to run at a higher pace for quite a while after the session has finished. Exercise also helps keep the muscles – and you – in shape which is just as important as watching the weight on the scales.

Another part of balancing the diet is not over-doing the alcohol. A couple of glasses of wine a day is around the limit to set yourself. There are many delicious non-alcoholic drinks and de-alcoholised wines and beers. After all, alcohol, like sugar, is 'empty calories' which means it gives you no nutrients (like vitamins and minerals), just calories. It robs the body of nutrients to digest it – nutrients that are needed to run vital bodily functions and protect us from infections and colds.

Here are some suggestions which will help you plan a healthier day's eating and illustrate the versatility of oats.

**Breakfast**

Porridge
Muesli – based on oats with either/or barley flakes, millet, sunflower seeds, sesame seeds, pumpkin seeds, dried fruits, nuts – served with natural yoghurt, skimmed milk or fruit juice and mineral water mixed
Oat muffins
Bran and oat muffins
Granola cereal with oats and dried fruits
Pancakes – oat or wholemeal
Waffles – oat or wholemeal
Breakfast scones – oat and wholemeal
Wholemeal bread or toast – thinly spread with soft vegetable margarine and no added sugar jams, marmalades, honey, yeast extract
Dried fruit compõté
Fresh fruits
Natural yoghurt

Free-range egg, boiled or scrambled (without fat or salt)
Baked beans
Brown rice kedgeree
Mushrooms on toast – cooked in bouillon, not fried
Milk – skimmed or try goats milk
Mineral water
Fresh fruit juices
Herb teas
Decaffeinated coffee

**Desserts**

Natural yoghurt – on its own or with fresh fruit or honey
Fresh fruit – salads, platters
Crumbles – fruit base with wholemeal and or oat crumble topping
Pancakes – fruit fillings or lemon and honey toppings
Ice-cream – topped with toasted oatmeal
Fools – made with fresh or dried fruits with crunchy oat topping
Cheesecakes – low fat cheeses with crunchy biscuit bases
Fruit tarts – wholemeal and oat pastry bases

**Snacks and bakes**

Oatmeal breads
Tea breads
High fibre cakes
Oaty crunchy bars
Slices – apricot and date
Oatcakes with a little cheese
Savoury scones
Fresh fruit
Dried fruit
Nuts
Yoghurt
Sandwiches – toasted or plain
Soups
Baked beans
Baked potatoes

**Lunch or Dinner**

Soups – vegetable, pulses, pastas
Salads – raw or just-blanched vegetables
Fish – grilled or poached
Poultry or game – grilled or poached, skin removed

Quiches – wholemeal or oat pastry
Pies – wholemeal or oat pastry
Crumbles – savoury mixtures of vegetables and beans topped with wholemeal flour, cheese and oats
Beans – dishes like chilli con carne, salads, casseroles
Brown rice – paellas, risotto, curry, stuffed vegetables, kedgeree, salads
Pasta – wholemeal or buckwheat
Pancakes – stuffed with vegetables, or meat or fish mixtures and made from wholemeal flour and oats
Herring and fishcakes – rolled in oats
Poultry – with oat stuffings for extra fibre
Sandwiches – wholemeal, oat breads
Open sandwiches – wholemeal bread, pumpernickel, rye crispbreads, rye breads, oatcakes topped with salad vegetables and fish or eggs.
Pizza – wholemeal or oat based
Rice salad
Wholemeal pasta salad
Quiches – wholemeal or oat pastries
Scones – savoury
Pitta breads – filled with salad and beans or other protein
Burgers – vegetable based
Salads – with root and green vegetables, plus beans or nuts and seeds, or cheese, fish, lean meat
Muesli – it's an anytime meal!

# *Chapter 5*
# OATS DOWN ON THE HEALTH FARM

Taking a few days out at a health farm or health hydro is an expensive business whether the resort is in this country or abroad. It is possible, with a little will power, to get the same benefits in the comfort of your own home. Obviously the health farm has the attraction of peace and quiet, no distractions and someone looking over your shoulder to make sure you don't break the rules (and sometimes there's an element of luxury, too). So, to make sure your Health Weekend is successful you will have to be a little disciplined.

Firstly, set aside two or three days when you will not have other arrangements or commitments to see friends or family. Having set aside these days in your diary you could contact a friend and see if he or she would like to join in. This way you can give each other moral support. You could always tell other friends that you are going away for the weekend so that no one will telephone to ask you out for a drink on the Saturday night!

This short health break is not a full-scale fast and even if it were there would be no need to treat yourself as an invalid, which is often a temptation, and stay in bed for the weekend! Moderate exercise is all part of the fun and fresh air will enhance the good effects of the three day diet plan.

Make sure you read through the diet first and do all your shopping etc. beforehand so that you can continue without any hitches or compromise meals.

**Friday**

At work today you could cut down, or cut out, regular tea and coffee and drink instead mineral water and/or fruit and vegetable juices. Omit the Friday lunchtime visit to the pub and have just a wholemeal sandwich or salad for lunch.

When you get home you can relax and start the healthy break with a good mixed salad as your evening meal. Make sure it contains a good mixture of leaves and roots and leave out the onions and garlic for this meal. Make up your salad from any of the following: watercress, mustard and cress, sweetcorn, beetroot (raw and grated or cooked) tomato, carrots, celery, red or green peppers, courgettes, parsley, lettuce, white or red cabbage, baby turnips or parsnips, grated or shredded spinach, apples, sultanas, raisins, fennel.

Add to the salad: unsalted nuts of your choice, sunflower or pumpkin seeds, cooked whole wheat berries or brown rice, chick peas, beans or other pulses of your choice. Dress the salad with a little olive oil, freshly pressed lemon or orange juice, pinch of ready made meux style mustard. If you are very hungry you could eat some wholemeal bread with the salad, but it will be very filling and satisfying. Remember to take your time and chew well and relax afterwards.

To drink during the evening choose from mineral water, fruit and or vegetable juices, herb teas.

**Saturday**

If you feel hungry when you wake up start the day with home-made muesli. You can make this up the night before and leave it in the fridge to soak overnight in mineral water or a mixture of apple juice and water. Serve it with unsweetened natural yoghurt if you like the combination; otherwise eat it with fresh fruit sliced on top.

To make your muesli mix together rolled oats, millet flakes, flaked almonds or chopped hazelnuts, raisins or sultanas, sunflower seeds or pumpkin seeds. Cover with the juice and/or water and leave to soak overnight. Alternatively you could soak it in natural unsweetened yoghurt with a little lemon juice added, if liked. Top with fresh fruit in the morning or grate in a small eating apple which need not be peeled.

Again, remember to eat slowly, chew thoroughly and relax during and after your breakfast.

For lunch on Saturday a large salad, made from any of the ingredients outlined in Friday night's section, is on the menu, together with wholemeal bread, or, for a change, rye bread or a mixed grain loaf.

The evening meal on Saturday is a risotto made with brown rice. In this you can use any vegetables of your choice, but try to ensure they are a mixture of root and leaf vegetables. You can either boil the brown rice first and then stir in the raw, prepared vegetables or you could cook them lightly and stir into the cooked rice. Alternatively you may prefer to lightly sauté the diced or shredded vegetables in a little corn or soy oil, stir in the rice and then pour over boiling vegetable stock and leave the mixture to simmer for 35 minutes until the rice is cooked.

Desserts and snacks throughout. Saturday and Sunday should be fresh fruit. You might prefer cleansing fruits like grapes and pears rather than higher calorie bananas.

To drink during Saturday and Sunday choose from mineral water (still or sparkling) fruit and/or vegetable juices, herb teas and if you cannot go without coffee for the weekend use decaffeinated coffee. This is available as instant coffee, ground coffee or decaffeinated coffee beans.

**Sunday**

If you feel hungry on Sunday morning you could have muesli made as for Saturday's instructions, or you might prefer a lighter breakfast of fresh fruit and natural unsweetened yoghurt. Alternatively during the winter a dried fruit compôté, made from simmered prunes, dried apricots, dried apple rings, dried peaches and pears, is very good especially if simmered in water to which spices like cloves or cinnamon have been added. Serve with unsweetened natural yoghurt.

For Sunday lunch prepare a jacket potato and resist the temptation to smother it in butter – try instead filling it with cottage cheese or a low-fat quark flavoured with some freshly cut herbs.

Sunday's evening meal is a three bean salad made by boiling 50g (2 oz) of each of three beans of your choice such as borlotti, red or black kidney beans, flageolet, haricot or butter beans and then tossing the cooled cooked beans in the salad dressing (see Friday) and mixing with grated carrots, cabbage and celery, apple and a little cooked wholemeal pasta.

**Slimmers**

If you want to lose weight during the weekend you might to like leave out the lunches and eat only fresh fruit during the day but have the evening meal.

**Exercise**

If possible take a long brisk walk on both days in a park or in the countryside if possible. You could also take some other form of exercise like swimming or a gentle jog during the day. If you practise meditation or yoga these will enhance the benefits of the weekend as would even a simple keep-fit routine, especially if it was aerobically designed.

**Relaxation**

This could also be the weekend where you find time to read that book you have been meaning to read for months. Don't use it as a weekend to read company reports or to catch-up on office work! You might even get a cassette tape with meditation or relaxation technique instructions and put your spare time to some positive use.

# Soluble Fibre Table

| | Soluble | Insoluble | Total Dietary Fibre | Energy k calorie |
|---|---|---|---|---|
| | (Figures represent g/100g) | | | |
| **Oat-Based Cereals** (Sources of Beta Glucan) | | | | |
| Quaker Harvest Crunch – Apple & Bran | 3.86 | 5.20 | 9.06 | 416 |
| Quaker Oats/Scotts Porage Oats | 3.58 | 2.91 | 6.49 | 377 |
| Quaker Warm Start | 3.49 | 2.84 | 6.33 | 384 |
| Quaker Oat Krunchies | 3.06 | 2.52 | 5.58 | 370 |
| Quaker Hot Bran Cereal | 2.95 | 9.28 | 12.23 | 337 |
| **Other Cereals** | | | | |
| All Bran | 3.74 | 18.73 | 22.47 | 273 |
| Weetabix | 3.08 | 6.69 | 9.77 | 340 |
| Shredded Wheat | 2.13 | 7.69 | 9.82 | 324 |
| Rice Krispies | 0.19 | 0.66 | 0.85 | 372 |
| Kellogg's Cornflakes | 0.17 | 0.47 | 0.64 | 364 |
| **Bread/Flour** | | | | |
| Rye Flour | 3.99 | 7.86 | 11.85 | 335 |
| Brown Flour | 1.93 | 5.12 | 7.05 | 327 |
| White Flour | 1.39 | 0.90 | 2.29 | 337 |
| Brown Bread | 1.37 | 2.88 | 4.25 | 223 |
| White Bread | 1.16 | 0.44 | 1.60 | 233 |
| Cornflour | 0.06 | 0.04 | 0.10 | 354 |
| Semolina | 1.06 | 1.19 | 2.25 | 350 |
| **Rice and Pasta** | | | | |
| Spaghetti Wholewheat | 2.18 | 6.78 | 8.96 | 338 |
| Spaghetti | 1.67 | 1.05 | 2.72 | 370 |
| Macaroni | 1.40 | 1.22 | 2.62 | 370 |
| Brown Rice | 0.12 | 1.62 | 1.74 | 350 |
| White Rice | 0.11 | 0.39 | 0.50 | 360 |

| | Soluble | Insoluble | Total Dietary Fibre | Energy k calorie |
|---|---|---|---|---|
| | (Figures represent g/100g) | | | |
| **Vegetables/Pulses** | | | | |
| Haricot Beans | 6.39 | 8.72 | 15.11 | 271 |
| Cooked kidney beans | 2.80 | 3.30 | 6.10 | 272 |
| Baked Beans | 1.70 | 1.50 | 3.20 | 64 |
| Cabbage | 1.31 | 1.39 | 2.70 | 22 |
| Carrots | 1.09 | 1.21 | 2.30 | 23 |
| Brussels Sprouts | 0.80 | 2.52 | 3.32 | 26 |
| Runner Beans | 0.70 | 3.72 | 4.42 | 26 |
| Potatoes | 0.67 | 0.62 | 1.29 | 87 |
| Tomatoes | 0.65 | 1.27 | 1.68 | 14 |
| Lettuce | 0.38 | 0.55 | 0.93 | 12 |
| **Fruits/Nuts** | | | | |
| Hazelnuts | 1.45 | 2.88 | 4.32 | 380 |
| Peanuts | 0.90 | 5.10 | 6.00 | 570 |
| Apples | 0.72 | 1.21 | 1.93 | 46 |
| Raisins | 0.70 | 1.00 | 1.70 | 246 |
| Bananas | 0.61 | 0.50 | 1.11 | 79 |

Table compiled by the Oats Information Bureau.

## Oat Flour

Oat flour is used in several recipes. It can be made by simply placing oats or fine or medium oatmeal in a liquidiser or food processor and grinding to a flour. Keep a jar ready made up to add oat power to your cooking!

# Chapter 6
# BREAKFASTS

## COCONUT MUESLI

Serves 4 5g fibre/250 calories per portion.

***4 tablespoons rolled oats***
***2 tablespoons millet flakes***
***1 tablespoon sunflower seeds***
***1 tablespoon raisins***
***1 tablespoon flaked almonds***
***1 tablespoon ground coconut***
***1 tablespoon wheatgerm***

1. Mix all ingredients together very thoroughly to ensure fair shares of the nuts and raisins!
2. Place portions of muesli in breakfast bowls and soak overnight in either fruit juice (such as apple juice) or mineral water. Natural yoghurt or cultured buttermilk may also be used.

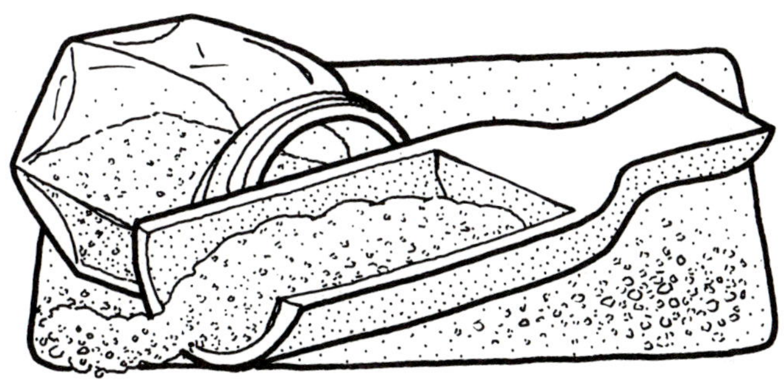

# DR ANDERSON'S OAT MUFFINS

Makes 24 2 g fibre/63 calories per muffin

| | *Metric* | *Imperial* | *American* |
|---|---|---|---|
| ***Oatbran and oatgerm*** | ***100 g*** | ***4 oz*** | ***1 cup*** |
| ***Oatflour*** | ***100 g*** | ***4 oz*** | ***1 cup*** |
| ***Wholemeal flour*** | ***225 g*** | ***8 oz*** | ***2 cups*** |
| ***Baking powder*** | ***2 tsp*** | ***2 tsp*** | ***2 tsp*** |
| ***Skimmed milk*** | ***360 ml*** | ***12 fl oz*** | ***1¼ cups*** |
| ***Vegetable oil*** | ***2 tsp*** | ***2 tsp*** | ***2 tsp*** |
| ***Egg whites, whisked*** | ***2*** | ***2*** | ***2*** |

**Variations**

***Add either 50 g/2 oz/½ cup raisins, or 150 ml/5 fl oz/⅔ cup apple purée, or 1 diced pear.***

1. Lightly oil muffin or bun tins and heat the oven to 375°F/ 190°C (Gas Mark 5).
2. Sieve together the oatbran and the flours and baking powder. If using raisins stir them into the flour.
3. Mix together the milk and oil and add to the flour with the apple purée, raisins or diced pear if using.
4. Whisk the egg whites until stiff and fold into the mixture.
5. Spoon into muffin pans or individual paper cake cases and bake for about 25 minutes.

# RICH CROISSANTS

Makes 16 2 g fibre/190 calories per croissant

| | *Metric* | *Imperial* | *American* |
|---|---|---|---|
| ***Wholemeal flour*** | ***325 g*** | ***12 oz*** | ***3 cups*** |
| ***Oat flour*** | ***100 g*** | ***4 oz*** | ***1 cup*** |
| ***Evaporated milk*** | ***180 ml*** | ***6 fl oz*** | ***$\frac{3}{4}$ cup*** |
| ***Water*** | ***120 ml*** | ***4 fl oz*** | ***$\frac{1}{2}$ cup*** |
| ***Clear honey*** | ***1 tbsp*** | ***1 tbsp*** | ***1 tbsp*** |
| ***Fresh yeast*** | ***1 tbsp*** | ***1 tbsp*** | ***1 tbsp*** |
| ***Vitamin C tablet, crushed*** | | | |
| ***Unsalted butter*** | ***225 g*** | ***8 oz*** | ***1 cup*** |

1. Lightly oil 2 baking sheets and heat oven to 450°F/230°C (Gas Mark 8).
2. Sieve flours into mixing bowl. Put milk, water and honey in a saucepan over a low heat and warm. Remove from heat. Crumble in yeast and vitamin C and mix well.
3. Pour the mixture into the flour and mix to a dough. Cover and rest while flattening the butter into a rectangle. Use a floured rolling pin and work on a floured surface.
4. Roll the rested dough out into a large rectangle and place the butter on top. If difficult to handle slip a palette knife under the butter. Envelope the butter in the dough and fold into 3. Cover and rest in the fridge for 20 minutes.
5. After this period place the dough with the folded edge away from the body and roll out again. Fold into 3. Roll out and fold again. Cover the dough and rest in the fridge for 20 minutes. Repeat twice, refrigerating for 20 minutes between each rolling and folding. Cover and rest again for 15–20 minutes before use.
6. Divide the dough in half and roll each half into a large circle. Cut into 16 and roll up each croissant from the outside towards the centre.
7. Bend into a crescent shape and place on a lightly oiled baking tray.
8. Cover and leave to rise in a warm place until risen by $1\frac{1}{2}$–2 times original size. Glaze with lightly beaten egg or milk and bake for 20 minutes.

# Fruit and Nut Muesli

Serves 8 7.7 g fibre/330 calories per portion.

| | *Metric* | *Imperial* | *American* |
|---|---|---|---|
| ***Dried apricots*** | ***75 g*** | ***3 oz*** | ***¾ cup*** |
| ***Dates*** | ***75 g*** | ***3 oz*** | ***¾ cup*** |
| ***Almonds*** | ***75 g*** | ***3 oz*** | ***¾ cup*** |
| ***Walnuts*** | ***75 g*** | ***3 oz*** | ***¾ cup*** |
| ***Sunflower seeds*** | ***40 g*** | ***1½ oz*** | ***⅓ cup*** |
| ***Rolled oats*** | ***325 g*** | ***12 oz*** | ***3 cups*** |

1. Roughly chop the apricots and dates. Finely chop the almonds and walnuts. Toast the sunflower seeds and mix all ingredients with the oats. Stir together thoroughly and store in an airtight container until required.

# Fig Muffins

Makes 8 2 g fibre/118 calories per muffin

| | *Metric* | *Imperial* | *American* |
|---|---|---|---|
| ***Sesame or sunflower oil*** | ***3 tbsp*** | ***3 tbsp*** | ***3 tbsp*** |
| ***Skimmed milk*** | ***5 tbsp*** | ***5 tbsp*** | ***5 tbsp*** |
| ***Molasses*** | ***1 tbsp*** | ***1 tbsp*** | ***1 tbsp*** |
| ***Wholemeal flour*** | ***100 g*** | ***4 oz*** | ***1 cup*** |
| ***Baking powder*** | ***1 tsp*** | ***1 tsp*** | ***1 tsp*** |
| ***Rolled oats*** | ***50 g*** | ***2 oz*** | ***½ cup*** |
| ***Dried figs, finely chopped*** | ***50 g*** | ***2 oz*** | ***½ cup*** |
| ***Free-range egg*** | ***1*** | ***1*** | ***1*** |

1. Lightly oil a muffin pan or bun tin (for 8 muffins) and heat oven to 375°F/190°C (Gas Mark 5).
2. Place the oil, 3 tablespoons of milk and the molasses in a saucepan and place over a low heat to melt the mixture. Remove from the heat.
3. Sieve the flour into mixing bowl with the baking powder and stir in the oats. Prepare the figs and stir into the flour.

4. Pour in the oil mixture. Lightly beat the egg with the rest of the milk and add to the mixture.
5. Place tablespoonfuls of the mixture into the tin and bake for 20–25 minutes until firm and golden brown.

# Wheatgerm Muffins

Makes 12 2 g fibre/140 calories per muffin

| | *Metric* | *Imperial* | *American* |
|---|---|---|---|
| ***Sesame oil*** | ***3 tbsp*** | ***3 tbsp*** | ***3 tbsp*** |
| ***Honey*** | ***2 tbsp*** | ***2 tbsp*** | ***2 tbsp*** |
| ***Molasses*** | ***1 tbsp*** | ***1 tbsp*** | ***1 tbsp*** |
| ***Wholemeal flour*** | ***100 g*** | ***4 oz*** | ***1 cup*** |
| ***Wheatgerm*** | ***50 g*** | ***2 oz*** | ***½ cup*** |
| ***Oats*** | ***50 g*** | ***2 oz*** | ***½ cup*** |
| ***Baking powder*** | ***3 tsp*** | ***3 tsp*** | ***3 tsp*** |
| ***Free-range egg*** | ***1*** | ***1*** | ***1*** |
| ***Sultanas*** | ***100 g*** | ***4 oz*** | ***1 cup*** |
| ***Walnuts, chopped*** | ***50 g*** | ***2 oz*** | ***½ cup*** |
| ***Apple, grated*** | ***1*** | ***1*** | ***1*** |
| ***Water*** | ***6 tbsp*** | ***6 tbsp*** | ***6 tbsp*** |

1. Lightly oil a muffin pan or bun tin (for 12 muffins) and heat oven to 375°F/190°C (Gas Mark 5).
2. Warm oil, honey and molasses in a saucepan.
3. Place rest of ingredients, except water, in a mixing bowl. Pour on the melted mixture and stir well. Add enough water to give a soft dropping consistency.
4. Place generous spoonfuls into the tin and bake for 20–30 minutes.

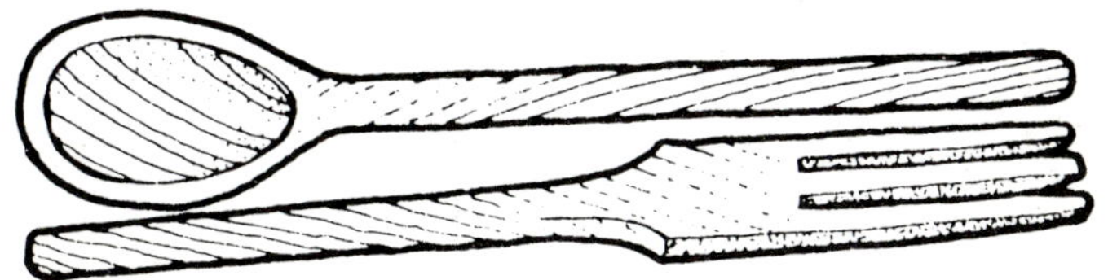

# Soft Oat Muffins

Makes 16 muffins    1 g fibre/70 calories per muffin

| | *Metric* | *Imperial* | *American* |
|---|---|---|---|
| ***Oats*** | ***100 g*** | ***4 oz*** | ***1 cup*** |
| ***Water*** | ***300 ml*** | ***$\frac{1}{2}$ pint*** | ***$1\frac{1}{4}$ cups*** |
| ***Wholemeal flour*** | ***50 g*** | ***2 oz*** | ***$\frac{1}{2}$ cup*** |
| ***Baking powder*** | ***2 tsp*** | ***2 tsp*** | ***2 tsp*** |
| ***Vegetable oil*** | ***2 tbsp*** | ***2 tbsp*** | ***2 tbsp*** |
| ***Molasses*** | ***1 tbsp*** | ***1 tbsp*** | ***1 tbsp*** |
| ***Sultanas*** | ***75 g*** | ***3 oz*** | ***$\frac{3}{4}$ cup*** |
| ***Apple, grated*** | ***1*** | ***1*** | ***1*** |
| ***Free-range egg, beaten*** | ***1*** | ***1*** | ***1*** |

1. Lightly oil a bun tray (or use paper baking cases) and heat oven to 375°F/190°C (Gas Mark 5).
2. Place oats in saucepan with water and gradually make a porridge by stirring over a low heat. When the mixture has thickened remove from heat.
3. Sieve flour into mixing bowl with baking powder. Place oil, molasses and sultanas in a saucepan over a low heat until the first two are melted together.
4. Add cooled porridge to flour and stir well. Stir in the molasses mixture and add the grated apple or carrot. Stir in beaten egg.
5. Place generous tablespoonfuls into prepared tins. Bake for about 25 minutes.

# Hot Waffles

Waffles make a delicious hot breakfast, similar to pancakes, but less fuss. Top with natural yoghurt, fresh fruit, fruit purées, stewed fruit, or a little honey.

Makes 8    1.5 g fibre/135 calories per waffle

| | Metric | Imperial | American |
|---|---|---|---|
| *Wholemeal flour* | *100 g* | *4 oz* | *1 cup* |
| *Oat flour* | *50 g* | *2 oz* | *½ cup* |
| *Baking powder* | *2 tsp* | *2 tsp* | *2 tsp* |
| *Free-range eggs, separated* | *2* | *2* | *2* |
| *Skimmed milk* | *300 ml* | *½ pint* | *1¼ cups* |
| *Soft vegetable margarine, melted* | *50 g* | *2 oz* | *¼ cup* |

1. Sieve the flours and baking powder into a mixing bowl.
2. Lightly beat the egg yolks and place in a well in the centre of the flour with a little of the milk and whisk with a fork, gradually drawing in the flour.
3. As the mixture thickens add more milk to the batter.
4. Whisk the egg whites until stiff. Carefully drizzle the melted margarine into the batter. Fold in the egg whites.
5. Heat the waffle maker and add 2 or 3 tablespoonfuls of batter. Close the lid and cook for 1–2 minutes until golden brown. Serve hot.

# Corn Muffins

Makes 12 0.5 g fibre/75 calories each

| | Metric | Imperial | American |
|---|---|---|---|
| *Free-range egg, separated* | *1* | *1* | *1* |
| *Free-range egg yolk* | *1* | *1* | *1* |
| *Corn oil* | *2 tbsp* | *2 tbsp* | *2 tbsp* |
| *Milk* | *2 tbsp* | *2 tbsp* | *2 tbsp* |
| *Corn/maize meal* | *100 g* | *4 oz* | *1 cup* |
| *Oat flour* | *50 g* | *2 oz* | *½ cup* |
| *Baking powder* | *1 tsp* | *1 tsp* | *1 tsp* |

1. Lightly oil a muffin pan or bun tin(s) with space for 12 buns and set oven to 400°F/200°C (Gas Mark 6).
2. Beat together the egg yolks, oil and milk. Sieve the corn/maize meal, oat flour and baking powder into a mixing bowl.

# CORN MUFFINS continued

3. Mix in the liquid and spoon the mixture into the prepared baking tins.
4. Bake for 20 minutes in the centre of the oven. Remove and allow to cool slightly before removing from the pan or bun tin and serving warm. Delicious for breakfast with no added sugar jams.

# Plain Porridge

Serves 1  7g fibre/100 calories

1. For a simple bowl of porridge for any time of the day or night place a handful of oats (which is about 2 tablespoons) in a saucepan with 150 ml/$\frac{1}{4}$ pint/$\frac{2}{3}$ cup water or skimmed milk and stir over a moderate heat until thick and creamy. This takes about 3–5 minutes.

# BREAKFAST SCONES

Makes 8 2.5 g fibre/150 calories per scone

| | *Metric* | *Imperial* | *American* |
|---|---|---|---|
| ***Muesli (no-added-sugar)*** | ***100 g*** | ***4 oz*** | ***1 cup*** |
| ***Water*** | ***150 ml*** | ***$\frac{3}{4}$ pint*** | ***$\frac{2}{3}$ cup*** |
| ***Wholemeal flour*** | ***100 g*** | ***4 oz*** | ***1 cup*** |
| ***Baking powder*** | ***2 tsp*** | ***2 tsp*** | ***2 tsp*** |
| ***Unsalted butter*** | ***50 g*** | ***2 oz*** | ***$\frac{3}{4}$ cup*** |
| ***Sultanas*** | ***50 g*** | ***2 oz*** | ***$\frac{1}{2}$ cup*** |
| ***Milk or beaten egg to glaze*** | | | |

1. Lightly oil a baking tray and heat oven to 450°F/230°C (Gas Mark 8).
2. Place the muesli in a cereal bowl and stir in the water. Leave to soak until the grains have become swollen (about 30 minutes or overnight in the fridge).
3. Sieve the flour into a mixing bowl with the baking powder. Add the butter and rub in until the mixture resembles breadcrumbs in consistency. Stir in the sultanas.
4. Add the soaked muesli and stir to a soft dough. Turn on to a lightly floured surface and knead a little until the dough is smooth and holds its shape.
5. Place in an 18cm (7 inch) round on the baking tray and score the top with a sharp knife into 8 sections. Glaze with a little milk or beaten egg and bake for 20 minutes.

# Fruity Porridge

Serves 4 4.5 g fibre/240 calories per portion

| | *Metric* | *Imperial* | *American* |
|---|---|---|---|
| ***Rolled oats*** | ***175 g*** | ***6 oz*** | ***1½ cups*** |
| ***Wheatgerm*** | ***50 g*** | ***2 oz*** | ***¼ cup*** |
| ***Raisins*** | ***50 g*** | ***2 oz*** | ***½ cup*** |
| ***Water*** | | | |
| ***Eating apples*** | ***2*** | ***2*** | ***2*** |

1. Place the oats, wheatgerm and raisins in a saucepan.
2. Add water and place over low heat. Bring to simmering point stirring continuously to prevent sticking. The amount of water needed will depend on the absorbency of the oats and the texture at which porridge is preferred. Between ½ pint (350 ml) and ¾ pint (450 ml) will probably be enough.
3. Core, but do not peel the apples, and grate them into the porridge just before serving.

# Family Favourite Granola

Makes 14 servings 3 g fibre/190 calories per serving

| | *Metric* | *Imperial* | *American* |
|---|---|---|---|
| ***Sesame, corn oil or soy oil*** | ***4 tbsp*** | ***4 tbsp*** | ***4 tbsp*** |
| ***Clear honey*** | ***2 tbsp*** | ***2 tbsp*** | ***2 tbsp*** |
| ***Natural vanilla essence*** | ***3 drops*** | ***3 drops*** | ***3 drops*** |
| ***Rolled oats*** | ***225 g*** | ***8 oz*** | ***2 cups*** |
| ***Sunflower seeds*** | ***50 g*** | ***2 oz*** | ***½ cup*** |
| ***Hazelnuts (half whole/half chopped)*** | ***50 g*** | ***2 oz*** | ***½ cup*** |
| ***Flaked almonds*** | ***50 g*** | ***2 oz*** | ***½ cup*** |
| ***Desiccated coconut*** | ***50 g*** | ***2 oz*** | ***½ cup*** |
| ***Wheatgerm*** | ***50 g*** | ***2 oz*** | ***¼ cup*** |
| ***Raisins or sultanas*** | ***50 g*** | ***2 oz*** | ***½ cup*** |

# GRANOLA continued

1. Lightly oil a large shallow baking tin and heat oven to 325°F/190°C (Gas Mark 5).
2. Place oil and honey in a saucepan over a low heat and melt together. Remove from heat and add vanilla.
3. Place rest of ingredients, except raisins, in a mixing bowl and pour on the oil. Stir thoroughly.
4. Place in prepared tin and bake for 30 minutes. Every 5–7 minutes stir and turn the mixture to ensure it is evenly toasted. Remove from oven.
5. Cool and stir in raisins. When completely cold store in an airtight container.

# Easy Croissants

Makes 12 2 g fibre/185 calories per croissant

| | *Metric* | *Imperial* | *American* |
|---|---|---|---|
| ***Wholemeal flour*** | ***225 g*** | ***8 oz*** | ***2 cups*** |
| ***Oat flour*** | ***100 g*** | ***4 oz*** | ***1 cup*** |
| ***Unsalted butter*** | ***175 g*** | ***6 oz*** | ***$\frac{3}{4}$ cup*** |
| ***Skimmed milk*** | ***210 ml*** | ***7 fl oz*** | ***1 cup*** |
| ***Fresh yeast*** | ***12 g*** | ***$\frac{1}{2}$ oz*** | ***1 tbsp*** |
| ***Clear honey*** | ***1 tbsp*** | ***1 tbsp*** | ***1 tbsp*** |
| ***Vitamin C tablet, crushed*** | ***1*** | ***1*** | ***1*** |

**1.** Lightly oil a baking sheet and heat oven to 450°F/230°C (Gas Mark 8).

2. Sieve flours into a mixing bowl and rub the butter in well but not as finely as when making pastry.
3. Heat the milk to 98°F/200°C. Crumble in yeast, vitamin C and stir in honey. Pour milk on to flour and beat well. Cover and leave to stand for 10 minutes.
4. Knead lightly on a floured surface and roll dough into a large circle. Cut into quarters and cut each quarter into 3 sections. Roll up each section from the outside towards the centre and bend into a crescent shape. Place on the tray, cover, and leave to double in size.
5. Glaze with milk or beaten egg and bake for 20 minutes.

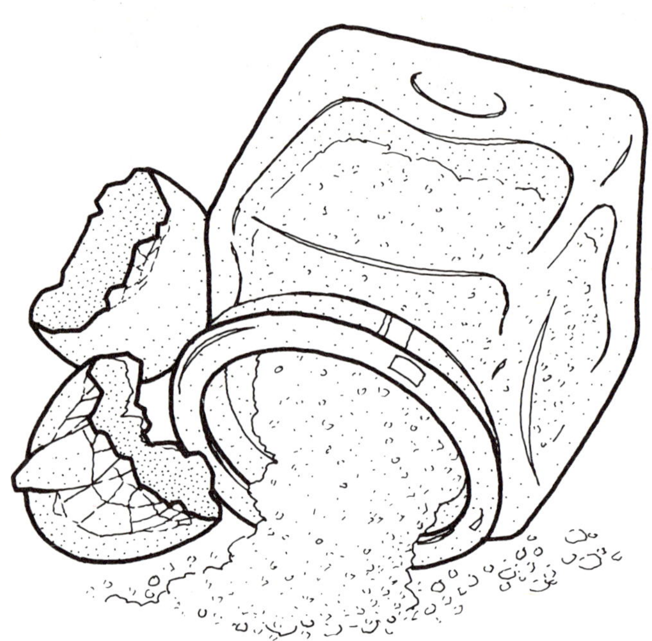

# Chapter 7
# MAIN MEALS

## SPINACH PANCAKES

Makes 10    6 g fibre/145 calories per pancake

| | *Metric* | *Imperial* | *American* |
|---|---|---|---|
| *Spinach, frozen, or* | *225 g* | *8 oz* | *2 cups* |
| *fresh spinach* | *450 g* | *11 b* | *4 cups* |
| *Oat flour* | *1 tbsp* | *1 tbsp* | *1 tbsp* |
| *Unsalted butter or soft vegetable margarine* | *1 tbsp* | *1 tbsp* | *1 tbsp* |
| *Nutmeg, fresh grated* | | | |
| *Pinch of sea salt* | | | |
| *Freshly ground black pepper* | | | |
| *Mature Cheddar cheese, grated* | *75 g* | *3 oz* | *¾ cup* |

1. Make the pancakes (page 48). Lightly oil an ovenproof dish and heat oven to 350°F/180°C (Gas Mark 4).
2. Wash fresh spinach in plenty of cold water to remove all dirt and grit. Place in a large saucepan and cover with a well-fitting lid. Cook for about 5 minutes, turning once or twice. (There is no need to add more water to cook the spinach in.) If using frozen spinach, drain well after defrosting, then heat gently.
3. Place the spinach in a liquidiser with the oat flour and butter and blend to a purée. Return to the saucepan and stir over a gentle heat while gradually adding the milk until a thick purée is achieved.
4. Season to taste and divide the mixture into equal amounts for each pancake. Roll up the pancakes and place in an ovenproof dish. Sprinkle cheese over the pancakes and bake for 20 minutes.

# Chicken Chervil Croquettes

Makes 16 0.5 g fibre/60 calories per croquette (including Oatcrunch Coating)

**Oatcrunch Coating**

| | *Metric* | *Imperial* | *American* |
|---|---|---|---|
| ***Wholemeal breadcrumbs*** | ***50 g*** | ***2 oz*** | ***½ cup*** |
| ***Rolled oats*** | ***25 g*** | ***1 oz*** | ***2 tbsp*** |
| ***Sesame seeds*** | ***25 g*** | ***1 oz*** | ***2 tbsp*** |

**1.** Mix together all the ingredients and place in a grill pan.

Toast, stirring from time to time to ensure even browning and prevent burning. This could also be done in a heavy-based frying pan which has been heated without fat.

**Croquettes**

| | Metric | Imperial | American |
|---|---|---|---|
| *4 chicken breasts, boneless* | *450 g* | *1 lb* | *1 lb* |
| *Medium or low-fat curd cheese* | *225 g* | *8 oz* | *2 cups* |
| *Lemon, rind and juice* | $\frac{1}{2}$ | $\frac{1}{2}$ | $\frac{1}{2}$ |
| *Chervil, freshly chopped* | *1 tbsp* | *1 tbsp* | *1 tbsp* |
| *Coarse grain ready-made mustard* | *1 tsp* | *1 tsp* | *1 tsp* |
| *Free-range egg whites* | 2 | 2 | 2 |

1. Lightly oil a baking sheet and heat the oven to 400°F/200°C (Gas Mark 6).
2. Mince the chicken and place in a bowl with the cheese, chervil, lemon rind, juice and mustard and beat to a smooth purée. This can be done in a food processor.
3. Whisk the egg whites until stiff and fold in 2 tablespoonfuls to lighten the mixture, then fold in the rest.
4. Mould into croquettes and roll in the Oatcrunch Coating. Place on the prepared tray and cook for 20 minutes, turning once.

# Eastern Fishcakes

Serves 4 0.5 g fibre/195 calories per portion

| | Metric | Imperial | American |
|---|---|---|---|
| *Mackerel, filleted and skinned* | *1* | *1* | *1* |
| *Coley or cod, filleted and skinned* | *175 g* | *6 oz* | *1$\frac{1}{2}$ cups* |
| *Potato, cooked and mashed* | *100 g* | *4 oz* | *1 cup* |
| *Clove garlic, crushed* | *1* | *1* | *1* |
| *Onion, grated* | *1* | *1* | *1* |
| *Free-range egg, lightly beaten* | *1* | *1* | *1* |
| *Freshly ground black pepper* | *1 tbsp* | *1 tbsp* | *1 tbsp* |

**Sauce**

| | Metric | Imperial | American |
|---|---|---|---|
| *Tomato purée* | *2 tbsp* | *2 tbsp* | *2 tbsp* |
| *Onion, grated* | *1* | *1* | *1* |
| *Red pepper, finely diced* | *1* | *1* | *1* |
| *Juice of 1 lemon* | | | |
| *Ground cumin* | *2 tsp* | *2 tsp* | *2 tsp* |
| *Red wine vinegar* | *2 tsp* | *2 tsp* | *2 tsp* |
| *Demerara sugar* | *1 tsp* | *1 tsp* | *1 tsp* |
| *Water* | *600 ml* | *1 pint* | *2½ cups* |
| *Oat/flour* | *1 tsp* | *1 tsp* | *1 tsp* |
| *Vegetable oil* | *2 tsp* | *2 tsp* | *2 tsp* |

1. Mince the prepared fish. Mix thoroughly with the remainder of the ingredients and form into 8 balls.
2. Smear a frying pan with oil and lightly sauté the fishcakes on all sides.
3. Mix all the ingredients for the sauce (except the oat flour).
4. Place the oat flour and oil in a saucepan and stir over a moderate heat to make a roux. Gradually stir in the sauce ingredients and after cooking for a couple of minutes carefully drop in the fishcakes.
5. Cover the pan and cook over a low heat for 30 minutes. Stir from time to time to keep the sauce smooth and prevent it sticking to the bottom of the pan.

# PIZZA SCONE

Serves 6 4.5 g fibre/315 calories per portion

| | Metric | Imperial | American |
|---|---|---|---|
| *Wholemeal flour* | *225 g* | *8 oz* | *2 cups* |
| *Oat flour* | *100 g* | *4 oz* | *1 cup* |
| *Baking powder* | *3 tsp* | *3 tsp* | *3 tsp* |
| *Soft vegetable margarine* | *75 g* | *3 oz* | *⅓ cup* |
| *Mustard powder* | *1 tsp* | *1 tsp* | *1 tsp* |
| *Gruyere cheese, grated* | *50 g* | *2 oz* | *½ cup* |
| *Cayene pepper* | *½ tsp* | *½ tsp* | *½ tsp* |
| *Lemon juice* | *2 tsp* | *2 tsp* | *2 tsp* |
| *Skimmed milk* | *150 ml* | *¼ pint* | *⅔ cup* |

**Topping**

| | *Metric* | *Imperial* | *American* |
|---|---|---|---|
| *Onion, diced* | *1* | *1* | *1* |
| *Green pepper, diced* | *1* | *1* | *1* |
| *Carrot, grated* | *1* | *1* | *1* |
| *Celery sticks, sliced* | *2* | *2* | *2* |
| *Corn or soya oil* | *1 tbsp* | *1 tbsp* | *1 tbsp* |
| *Canned tomatoes* | *400 g* | *14 oz* | *1¾ cups* |
| *Freshly ground black pepper* | | | |
| *Oregano* | *1 tsp* | *1 tsp* | *1 tsp* |
| *Basil* | *1 tsp* | *1 tsp* | *1 tsp* |
| *Potato flour or arrowroot* | | | |

1. Preheat the oven to 375°F/190°C (Gas Mark 5).
2. Sieve the flour and baking powder into a mixing bowl and rub in the fat until the mixture resembles fine breadcrumbs in consistency. Stir in the mustard, cheese and cayenne pepper. Add the lemon juice to the milk and stir into the flour mixture.
3. Turn on to a lightly floured surface and knead very lightly until the mixture forms a smooth dough. Roll out into a round and place on a baking tray, using a flan ring as a guide to make a 22 cm (7 inch) circle of dough. A saucer shape with a slightly raised edge will accommodate the topping more easily.
4. To make the pizza scone topping, place the prepared onion, pepper, carrot and celery in the oil in a heavy-based saucepan and cover. Sweat over a low heat for 5 minutes until transparent and softened.
5. Add the tomatoes, black pepper and herbs, and break up the tomatoes with the back of a spoon. Continue to cook for a further 10 minutes, uncovered so the mixture thickens. If it is still very liquid, thicken with a little potato flour or arrowroot slaked in water.
6. Spoon onto the pizza scone and bake for 20–25 minutes.

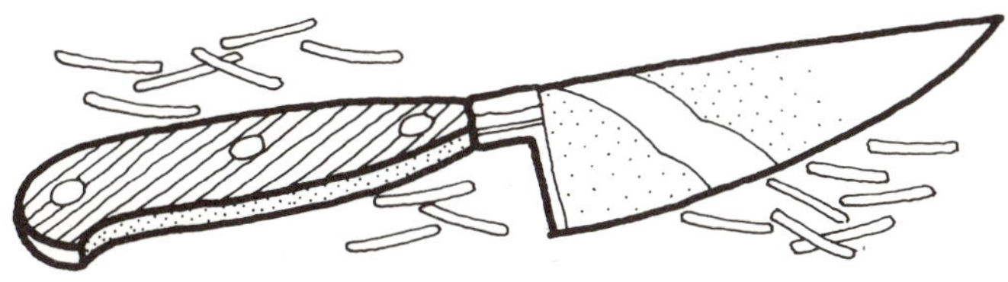

# Pancake Filling

Serves 4 3 g fibre/180 calories per pancake

| | *Metric* | *Imperial* | *American* |
|---|---|---|---|
| ***Courgette, large*** | ***1*** | ***1*** | ***1*** |
| ***Broccoli spears*** | ***2*** | ***2*** | ***2*** |
| ***Small aubergine*** | ***1*** | ***1*** | ***1*** |
| ***Leek*** | ***1*** | ***1*** | ***1*** |
| ***Chinese leaves*** | ***4*** | ***4*** | ***4*** |
| ***Carrots*** | ***2*** | ***2*** | ***2*** |
| ***Cashew nut pieces*** | ***100 g*** | ***4 oz*** | ***1 cup*** |
| ***Sesame oil*** | | | |
| ***Root ginger (optional)*** | | | |

1. Prepare the vegetables by washing well and cutting into matchstick strips, slices, or florets as appropriate.
2. Lightly oil a wok or large frying pan over a high heat. Toss in the vegetables in order of hardness and cook quickly, stirring all the time.
3. Just before serving toss in the cashew pieces and stir in a little freshly grated root ginger.
4. Stuff the pancakes with the vegetable mixture and pour over, or offer separately, the sweet and sour sauce.

# Sweet and Sour Sauce

Serves 4 0 g fibre/20 calories per serving

***1 tablespoonful tomato purée***
***2 tablespoonfuls shoyu (or soya) sauce***
***1 tablespoon wine, cider or sherry vinegar***
***2 tablespoonfuls water***
***½ tablespoonful demerara sugar***

1. Place all the ingredients in a saucepan and dissolve together over a low heat.
2. Thicken, if liked, with a little arrowroot or potato flour slaked in some cold water.

# Oat Pastry

13 g fibre/885 calories

Substitute one third wholemeal flour for rolled oats to give a pastry with a good rich texture.

| | *Metric* | *Imperial* | *American* |
|---|---|---|---|
| ***Wholemeal flour*** | ***113 g*** | ***4½ oz*** | ***1 cup*** |
| ***Rolled oats*** | ***50 g*** | ***2 oz*** | ***½ cup*** |
| ***Soft vegetable margarine or unsalted butter*** | ***75 g*** | ***3 oz*** | ***⅓ cup*** |
| ***Cold water to mix*** | | | |

1. Sieve the flour into a bowl, add the bran from the sieve and the rolled oats. Rub the fat in, until mixture resembles breadcrumbs in texture. Add a little cold water and mix to a soft, but not sticky dough. The pastry is now ready to use in savoury pies, pastries and flans.

**Pastry Case**

To prepare a pastry case roll out the Oat Pastry on a lightly floured surface, then line a 22 cm (8 inch) flan ring which has been lightly oiled.

To 'bake blind' (that is partly bake the pastry to seal it before adding a quiche or tart filling) place a layer of greaseproof paper on top of the pastry and fill with old beans. (These are usually called 'baking beans' and can be cooled and stored after use for future occasions.) The lined pastry case is then baked in a preheated oven at 400°F/200°C (Gas Mark 6) for 10 minutes. It is removed from the oven, the paper and beans are removed and it is now ready for filling. It can, of course, be completely cooked for 25 minutes – the last 5 without paper and beans, for use with an uncooked filling.

# Oat Pancakes

Makes 10    1 g fibre/50 calories per pancake

| | Metric | Imperial | American |
|---|---|---|---|
| *Wholemeal flour* | *50 g* | *2 oz* | *½ cup* |
| *Oat flour* | *50 g* | *2 oz* | *½ cup* |
| *Free-range egg, lightly beaten* | *1* | *1* | *1* |
| *Skimmed milk* | *300 ml* | *½ pint* | *1¼ cups* |

1. Sieve the flours together into a bowl. Make a well in the centre and add the egg. Using a fork gradually add the milk to the egg and work in the flour, until a smooth batter is achieved.
2. To make the pancakes lightly oil a 18 cm (7 inch) small heavy-based omelette pan and place over a high heat.
3. Using a small jug pour in enough batter to thinly coat the bottom of the pan, tipping the pan from side to side as you do it.
4. Cook for a minute then turn the pancake and cook the second side.

# Onion Tart

Serves 4    4.5 g fibre/359 calories per portion

| | Metric | Imperial | American |
|---|---|---|---|
| *Pastry case, baked blind* | *1* | *1* | *1* |
| *Onions, diced* | *2 large* | *2 large* | *2 large* |
| *Quark, or similar low-fat soft cheese* | *100 g* | *4 oz* | *1 cup* |
| *Natural yoghurt* | *150 ml* | *¼ pint* | *⅔ cup* |
| *Free-range eggs* | *2* | *2* | *2* |
| *Sea salt (optional)* | | | |
| *Freshly ground black pepper* | | | |
| *Pinch of ground mace* | | | |

1. Prepare the Oat Pastry flan case, page 47.

2. While the pastry case is baking place the onions in a lightly oiled frying pan and sauté for 5 minutes.
3. Beat together the cheese, yoghurt, eggs and seasoning.
4. Put the onions onto absorbent paper to drain any excess fat, and turn them into the prepared pastry case.
5. Pour the yoghurt mixture on top and return to the oven for 35 minutes or until golden brown and set.

# Spinach Soufflé

Serves 4 3 g fibre/140 calories per portion

| | *Metric* | *Imperial* | *American* |
|---|---|---|---|
| ***Soft vegetable margarine*** | ***50 g*** | ***2 oz*** | ***¼ cup*** |
| ***Oat flour*** | ***50 g*** | ***2 oz*** | ***1 cup*** |
| ***Skimmed milk*** | ***300 ml*** | ***½ pint*** | ***1¼ cups*** |
| ***Freshly ground nutmeg*** | | | |
| ***Freshly ground black pepper*** | | | |
| ***Gruyére cheese, grated*** | ***25 g*** | ***1 oz*** | ***2 tbsp*** |
| ***Spinach, cooked and chopped, or frozen*** | ***175 g*** | ***6 oz*** | ***1½ cups*** |
| ***Free-range eggs, separated*** | ***4*** | ***4*** | ***4*** |

1. Prepare an 18 cm (7 inch) soufflé dish by lightly oiling and making a double layered greaseproof paper collar which should overlap at least 2 cm (1 inch) at any joins and be 5 cm (2 inches) higher than the top of the dish. Tie in place with string. Heat the oven to 375°F/190°C (Gas Mark 5).
2. Place the fat and flour in a saucepan and stir over a low heat to make a roux. Gradually add the milk to make a thick sauce. Add the nutmeg and black pepper and remove from heat. Stir in the cheese and the spinach. Lightly beat the egg yolks and stir into the mixture.
3. Whisk the egg whites until stiff and fold in 2 tablespoonfuls to lighten the mixture. Fold in the remainder and quickly pour into the prepared soufflé dish.
4. Cook for 40 minutes until risen and set. Serve immediately with a lightly tossed green salad.

# Spinach Roulade

Serves 8 2 g fibre/195 calories per serving

| | *Metric* | *Imperial* | *American* |
|---|---|---|---|
| ***Soft vegetable margarine*** | ***75 g*** | ***3 oz*** | ***$\frac{1}{3}$ cup*** |
| ***Wholemeal flour*** | ***50 g*** | ***2 oz*** | ***$\frac{1}{2}$ cup*** |
| ***Oat flour*** | ***50 g*** | ***2 oz*** | ***$\frac{1}{2}$ cup*** |
| ***Skimmed milk*** | ***600 ml*** | ***1 pint*** | ***$2\frac{1}{2}$ cups*** |
| ***Free-range eggs, separated*** | ***3*** | ***3*** | ***3*** |
| ***Parmesan cheese*** | ***2 tbsp*** | ***2 tbsp*** | ***2 tbsp*** |
| ***Frozen spinach, thawed, or*** | ***175 g*** | ***6 oz*** | ***$1\frac{1}{2}$ cups*** |
| ***Fresh spinach*** | ***450 g*** | ***1 lb*** | ***1 lb*** |
| ***Shelled prawns*** | ***175 g*** | ***6 oz*** | ***$1\frac{1}{2}$ cups*** |

1. Line a small Swiss roll tin with greaseproof paper and set the oven to 350°F/180°C (Gas Mark 4).
2. Place the margarine and flour in a saucepan and stir over a moderate heat to make a roux. Gradually add the milk to the paste, stirring all the time to prevent lumps forming.
3. Remove from the heat and beat in the egg yolks, cheese and spinach. (If using fresh spinach wash well and place in a large saucepan over a moderate heat with the lid on. Cook for a few minutes, turning once or twice, drain and chop finely.)
4. Whisk the egg whites until stiff and fold into the spinach mixture. Pour into the prepared tin and bake for 20–25 minutes until an inserted skewer comes out clean.
5. Remove from the oven and invert on to a clean sheet of greaseproof paper. Cover for a minute with a cold, wet teatowel (to make the paper come off more easily).
6. Peel off the paper and sprinkle with the prawns. Trim the edges and roll up, using the clean paper to hold the roulade together. Serve at once with wedges of lemon.

# Spinach and Yoghurt Soup

Serves 6 3 g fibre/80 calories per serving

| | *Metric* | *Imperial* | *American* |
|---|---|---|---|
| ***Large onion, chopped*** | ***1*** | ***1*** | ***1*** |
| ***Clove garlic, crushed*** | ***1*** | ***1*** | ***1*** |
| ***Fresh spinach*** | ***450 g*** | ***1 lb*** | ***4 cups*** |
| ***or frozen spinach*** | ***300 g*** | ***10 oz*** | ***2½ cups*** |
| ***Stock*** | ***900 ml*** | ***1½*** | ***3¾ cups*** |
| ***Oat flour*** | ***50 g*** | ***2 oz*** | ***½ cup*** |
| ***Skimmed milk*** | | | |
| ***Nutmeg*** | | | |
| ***Freshly ground black pepper*** | | | |
| ***Sea salt*** | | | |
| ***Free-range egg yolk*** | ***1*** | ***1*** | ***1*** |
| ***Natural yoghurt*** | ***150 ml*** | ***¼ pint*** | ***⅔ cup*** |

1. Sauté the onion and garlic in a little oil until softened but not browned. Add the spinach and stock and cook with the lid on for 10 minutes.
2. Purée the soup in a liquidiser and return to the pan.
3. Mix the oat flour with the milk and add to the pan with the nutmeg and seasonings. Stir while thickening, bring to the boil and simmer for 5 minutes.
4. Remove from the heat and add the egg yolk mixed with the yoghurt.
5. Serve the soup with an extra swirl of yoghurt and toasted croutons for special occasions.

# Leek Pie

Serves 4 5.5 g/280 calories per serving

**Filling**

| | *Metric* | *Imperial* | *American* |
|---|---|---|---|
| *Leeks* | *450 g* | *1 lb* | *4 cups* |
| *Free-range eggs* | *2* | *2* | *2* |
| *Mature Cheddar cheese, grated* | *50 g* | *2 oz* | *½ cup* |
| *Natural yoghurt* | *150 ml* | *¼ pint* | *⅔ cup* |
| *Freshly ground black pepper* | | | |
| *Sea salt* | | | |

**Pastry**

| | | | |
|---|---|---|---|
| *Wholemeal flour* | *75 g* | *3 oz* | *¾ cup* |
| *Porridge oats* | *25 g* | *1 oz* | *2 tbsp* |
| *Soft vegetable margarine* | *50 g* | *2 oz* | *¼ cup* |
| *Sea salt* | | | |
| *Cold water to mix* | | | |

1. Lightly oil a pie dish and set the oven to 400°F/200°C (Gas Mark 6).
2. Trim leeks leaving 2 cm (1 inch) green leaves at top, and cut off roots. Cut into 2 cm (1 inch) slices and wash thoroughly. Plunge into boiling water or lightly steam for 5 minutes. Drain and place in base of deep pie dish.

3. Beat eggs with cheese, yoghurt, and seasoning. Pour over leeks.
4. Make pastry by sieving flour into mixing bowl. Stir in the oats and rub in margarine until mixture resembles breadcrumbs. Stir in pinch of sea salt. Add sufficient cold water to mix to a soft dough. Turn on to a lightly floured surface and roll out to size of pie dish leaving an extra 2 cm (1 inch) all round.
5. Cut out fine strip of pastry and place on rim of pie dish. Brush with water then carefully lift the pastry over the pie dish and place on top. Trim with a knife. Knock edges of pastry together with the back of a knife and flute. Roll out pastry trimmings to make some decorative leaves and place these in centre of pie.
6. Glaze with milk or a little beaten free-range egg and bake for 30 minutes until golden brown.

# WALNUT MOUSSAKA

Serves 6 4 g fibre/170 calories per serving

| | *Metric* | *Imperial* | *American* |
|---|---|---|---|
| *Onions, chopped* | *2* | *2* | *2* |
| *Mushrooms, sliced* | *100 g* | *4 oz* | *1 cup* |
| *Oats* | *50 g* | *2 oz* | *½ cup* |
| *Walnuts* | *100 g* | *4 oz* | *1 cup* |
| *Wholemeal breadcrumbs* | *50 g* | *2 oz* | *½ cup* |
| *Freshly ground black pepper* | | | |
| *Sea salt* | | | |
| *Canned tomatoes* | *400 g* | *14 oz* | *3½ cups* |
| *Aubergines* | *2 medium* | *2 medium* | *2 medium* |

**Topping**

| | | | |
|---|---|---|---|
| *Free-range egg* | *1* | *1* | *1* |
| *Oat flour* | *1 tbsp* | *1 tbsp* | *1 tbsp* |
| *Natural yoghurt* | *150 ml* | *¼ pint* | *⅔ cup* |
| *Grated cheese* | *50 g* | *2 oz* | *½ cup* |
| *Sea salt* | | | |
| *Freshly ground black pepper* | | | |

# Walnut Moussaka contd

1. Lightly oil a rectangular ovenproof dish and heat oven to 375°F/190°C (Gas Mark 5).
2. Sauté the onions in a little oil until softened but not browned. Add the mushrooms and cook for 1 minute.
3. Add the oats, walnuts and breadcrumbs, mix well and season with plenty of black pepper and a little sea salt.
4. Slice the aubergines and sprinkle with salt to bring out any bitter juices. Leave for 30 minutes then rinse thoroughly.
5. Place one third of the aubergine in a layer in the bottom of the dish. Place half the walnut mixture on top and add the tomatoes. Add another third of the aubergines, the remaining walnut mixture and the last of the aubergines.
6. Mix together the ingredients for the topping and spread over. Cook for 30 minutes.

# Mushroom Fish Pie

Serves 4 1.5 g fibre/370 calories per serving

**Basic Oat Pastry**, page 47.

**Filling**

| | Metric | Imperial | American |
|---|---|---|---|
| *Large onion, diced* | *1* | *1* | *1* |
| *Oil* | *1 tbsp* | *1 tbsp* | *1 tbsp* |
| *Mushrooms, sliced* | *175 g* | *6 oz* | *1½ cups* |
| *Oat flour* | *1 tbsp* | *1 tbsp* | *1 tbsp* |
| *Skimmed milk* | *150 ml* | *¼ pint* | *⅔ cup* |
| *Cod, or other white fish* | *450 g* | *1 lb* | *1 lb* |
| *Freshly ground black pepper* | | | |

1. Make the pastry and heat oven to 400°F/200°C (Gas Mark 6).
2. Sauté the onion in the oil until transparent, and stir in the mushrooms. Cover and cook for 5 minutes. Stir in the flour and gradually add the milk to form a thick sauce.
3. Trim the skin from the fish and remove bones. Cut into strips or chunks and add to the mushrooms. Season with pepper. Cook for 2 minutes. Then turn into an ovenproof dish.
4. Roll out the pastry to the circumference of the dish. Cut strips from the excess and place around the rim. Brush with water and place the pastry lid on top. Knock up the edges of the pastry to seal with the strip, decorate with shapes cut from the trimmings. Glaze with milk or beaten egg and bake for 30 minutes.

# PARSNIP CROQUETTES

Makes 8 3.5 g fibre/120 calories per croquette

| | *Metric* | *Imperial* | *American* |
|---|---|---|---|
| ***Parsnips*** | ***1 kg*** | ***2 lb*** | ***2 lb*** |
| ***Unsalted butter or soft vegetable margarine*** | ***50 g*** | ***2 oz*** | ***¼ cup*** |
| ***Wholemeal breadcrumbs*** | ***100 g*** | ***4 oz*** | ***1 cup*** |
| ***Mature cheese, grated*** | ***100 g*** | ***4 oz*** | ***1 cup*** |
| ***Rolled oats*** | ***50 g*** | ***2 oz*** | ***½ cup*** |
| ***Dry mustard powder*** | ***1 tsp*** | ***1 tsp*** | ***1 tsp*** |

1. Lightly oil a baking sheet and heat oven to 375°F/190°C (Gas Mark 5).
2. Scrub and/or peel parsnips and roughly chop, removing any very woody central core. Steam or boil until soft. Drain and mash or liquidise with the butter and most of the cheese. Stir in the breadcrumbs. When cool enough to handle form into croquettes.
3. Mix together the oats, remainder of the cheese and mustard and roll the croquettes in the mixture.
4. Place on baking tray and cook for 20 minutes, turning once.

# Braised Meatballs and Vegetables

Serves 6 1.5 g fibre/95 calories per portion

**Meatballs**

| | Metric | Imperial | American |
|---|---|---|---|
| *Minced steak* | *450 g* | *1 lb* | *1 lb* |
| *Wholemeal breadcrumbs* | *25 g* | *1 oz* | *2 tbsp* |
| *Oat flour* | *25 g* | *1 oz* | *2 tbsp* |
| *Onion, minced or grated* | *1* | *1* | *1* |
| *Free-range egg* | *1* | *1* | *1* |
| *Allspice, ground* | *1 tsp* | *1 tsp* | *1 tsp* |
| *Freshly ground black pepper* | | | |

**Vegetable base**

| | Metric | Imperial | American |
|---|---|---|---|
| *Celery* | *6 sticks* | *6 sticks* | *6 sticks* |
| *Carrots* | *2* | *2* | *2* |
| *Onion* | *1 large* | *1 large* | *1 large* |
| *Vegetable oil* | *2 tsp* | *2 tsp* | *2 tsp* |
| *Vegetable stock* | *450 ml* | *$\frac{3}{4}$ pint* | *2 cups* |
| *Lemon juice* | *1 tsp* | *1 tsp* | *1 tsp* |
| *Freshly ground black pepper* | | | |

1. Heat the oven to 400°F/200°C (Gas Mark 6).
2. Stir together all of the meatball ingredients. Form into balls. Lightly sauté in a smear of oil.
3. Scrub the celery and cut into 2 cm (1 inch) pieces. Scrub the carrots and slice thinly. Finely dice the onion.
4. Sauté the vegetables in the oil for 3 minutes. Then place in dish and pour over the stock, lemon juice and seasoning.
5. Place the meatballs on top of the vegetables and stock and bake for 40 minutes.

# Mushroom and Cashew Lasagne

Serves 4 5 g fibre/375 calories per serving

| | *Metric* | *Imperial* | *American* |
|---|---|---|---|
| ***Onion, chopped*** | ***1 medium*** | ***1 medium*** | ***1 medium*** |
| ***Leeks, thinly sliced*** | ***225 g*** | ***8 oz*** | ***2 cups*** |
| ***Mushrooms, thinly sliced*** | ***225 g*** | ***8 oz*** | ***2 cups*** |
| ***Cashew nuts, chopped*** | ***100 g*** | ***4 oz*** | ***1 cup*** |
| ***Oats*** | ***50 g*** | ***2 oz*** | ***$\frac{1}{2}$ cup*** |
| ***Soy sauce*** | ***2 tsp*** | ***2 tsp*** | ***2 tsp*** |
| ***Fresh pasta, page 60*** | ***225 g*** | ***8 0z*** | ***2 cups*** |
| ***Canned, chopped tomatoes*** | ***400 g*** | ***14 oz*** | ***$3\frac{1}{2}$ cups*** |

| | | | |
|---|---|---|---|
| *White sauce, made with oat flour* | *300 ml* | *½ pint* | *1½ cups* |
| *Low fat soft cheese* | *100 g* | *4 oz* | *½ cup* |
| *Parsley, chopped* | *1 tbsp* | *1 tbsp* | *1 tbsp* |

1. Lightly oil a shallow rectangular ovenproof dish and set oven to 375°F/190°C (Gas Mark 5).
2. Sauté the onion and leeks in a little oil until softened but not browned. Add the mushrooms, nuts, oats and soy sauce. Mix thoroughly and remove from heat.
3. Bring 1.5 litres (3 pints) water to the boil in a large saucepan. Add 1 tablespoonful of oil to prevent pasta sticking, then add a few squares of homemade pasta at a time. Cook for approximately 6 minutes. (If using bought dried pasta, cook for 10–12 minutes.) Drain.
4. Mix together the tomatoes, white sauce and cheese.
5. Place half the cashew sauce in the dish and cover with a layer of pasta, then a layer of tomato and cheese sauce. Repeat, ending with a layer of sauce.
6. Bake for 30 minutes. Serve sprinkled with a little chopped parsley.

# SPICED CRISPY CHICKEN

Serves 4 1.5 g fibre/525 calories per serving

| | *Metric* | *Imperial* | *American* |
|---|---|---|---|
| *Chicken legs* | *4* | *4* | *4* |
| *Free-range egg, beaten* | *1* | *1* | *1* |
| *Oats* | *100 g* | *4 oz* | *1 cup* |
| *Celery salt* | *1 tsp* | *1 tsp* | *1 tsp* |
| *Garlic powder (to taste)* | | | |
| *Paprika pepper* | *½ tsp* | *½ tsp* | *½ tsp* |

1. Lightly oil an ovenproof dish and set oven to 375°F/190°C (Gas Mark 5).
2. Trim and wash chicken. Remove the skin if liked. Coat in beaten egg and then in oats mixed with seasonings. Sprinkle any extra oats over the top.
3. Bake for 1 hour.

# HOMEMADE PASTA

26 g fibre/1160 calories

| | Metric | Imperial | American |
|---|---|---|---|
| *Wholemeal flour* | *225 g* | *8 oz* | *2 cups* |
| *Oat flour* | *100 g* | *4 oz* | *1 cup* |
| *Free-range eggs* | *3* | *3* | *3* |
| *Oil* | *1 tbsp* | *1 tbsp* | *1 tbsp* |
| *Water* | *2 tbsp* | *2 tbsp* | *2 tbsp* |
| *Fine sea salt* | *1 tsp* | *1 tsp* | *1 tsp* |

1. Make a well in the centre of the flour, stir in the remaining ingredients and knead together to form a dough. Knead thoroughly, adding a little more flour if necessary.
2. Leave in a polythene bag to rest for 1 hour.
3. Divide the dough into 4. Pass it at least twice through a pasta machine or roll out thinly by hand and cut into appropriate shapes.

**Lasagne**

To make lasagne cut out 7 cm (3 inch) squares, or longer strips when the dough has been passed through the pasta machine or rolled out.

# PARSLEY SCONE DUMPLINGS

Makes 4 2 g fibre/155 calories per dumpling

| | Metric | Imperial | American |
|---|---|---|---|
| *Wholemeal flour* | *50 g* | *2 oz* | *1 cup* |
| *Oat flour* | *50 g* | *2 oz* | *1 cup* |
| *Baking powder* | *1½ tsp* | *1½ tsp* | *1½ tsp* |
| *Soft vegetable margarine* | *40 g* | *1½ oz* | *2½ tbsp* |
| *Freshly chopped parsley* | *1 tbsp* | *1 tbsp* | *1 tbsp* |
| *Freshly ground black pepper* | | | |
| *Cold water to mix* | | | |

1. Sieve flours and baking powder into mixing bowl. Rub in fat until mixture resembles breadcrumbs in consistency. Stir in parsley and pepper. Mix to a soft dough with cold water.
2. Form into 4 balls or roll out and cut into four thick scone shapes with cutters. The dumplings are now ready to be added to casseroles and stews.

# Beef Burgers

Makes 4 3 g fibre/260 calories per burger

| | *Metric* | *Imperial* | *American* |
|---|---|---|---|
| ***Lean steak, minced*** | ***325 g*** | ***12 oz*** | ***3 cups*** |
| ***Onion, finely chopped*** | ***1 medium*** | ***1 medium*** | ***1 medium*** |
| ***Carrot, grated*** | ***1 large*** | ***1 large*** | ***1 large*** |
| ***Tomato purée*** | ***1 tbsp*** | ***1 tbsp*** | ***1 tbsp*** |
| ***Thick slice wholemeal bread, grated to crumbs*** | ***1*** | ***1*** | ***1*** |
| ***Oat flour*** | ***2 tbsp*** | ***2 tbsp*** | ***2 tbsp*** |
| ***Ready-made mustard*** | ***2 tbsp*** | ***2 tbsp*** | ***2 tbsp*** |

1. Place all ingredients in a mixing bowl and combine well.
2. Shape into 4 burgers.
3. Place under the grill and grill for 10 minutes on each side.
4. Serve at once with green salad.

# Swede and Carrot Gratin

Serves 6 4.5 g fibre/165 calories per serving

| | *Metric* | *Imperial* | *American* |
|---|---|---|---|
| ***Carrots*** | ***450 g*** | ***1 lb*** | ***1 lb*** |
| ***Swede*** | ***450 g*** | ***1 lb*** | ***1 lb*** |
| ***Soft vegetable margarine*** | ***25 g*** | ***1 oz*** | ***$\frac{3}{4}$ cup*** |
| ***Milk skimmed*** | ***60 ml*** | ***2 fl oz*** | ***$\frac{1}{4}$ cup*** |
| ***Pinch sea salt*** | | | |

| | | | |
|---|---|---|---|
| *Freshly ground black pepper* | | | |
| *Free-range eggs* | *2* | *2* | *2* |
| *Oats* | *75 g* | *3 oz* | *¾ cup* |
| *Mature Cheddar cheese, grated* | *50 g* | *2 oz* | *½ cup* |

1. Lightly oil an ovenproof dish and heat oven to 375°F/190°C (Gas Mark 5).
2. Peel and roughly chop the carrots and swede and steam or boil until just cooked. Liquidise or mash with fat, milk, seasoning and eggs. Place in the dish. Mix oats and cheese and sprinkle on top.
3. Bake for 25 minutes and serve at once.

# Vegetable Burgers

Makes 6 5 g fibre/330 calories per burger

| | *Metric* | *Imperial* | *American* |
|---|---|---|---|
| *Ground walnuts* | *100 g* | *4 oz* | *1 cup* |
| *Ground hazelnuts* | *100 g* | *4 oz* | *1 cup* |
| *Wholemeal breadcrumbs* | *100 g* | *4 oz* | *1 cup* |
| *Oat flour* | *50 g* | *2 oz* | *½ cup* |
| *Basil* | *1 tsp* | *1 tsp* | *1 tsp* |
| *Onion, diced* | *1 large* | *1 large* | *1 large* |
| *Free-range egg* | *1* | *1* | *1* |
| *Corn, soy or sesame oil* | *2 tbsp* | *2 tbsp* | *2 tbsp* |
| *Oats* | *2 tbsp* | *2 tbsp* | *2 tbsp* |
| *Sesame seeds* | *2 tbsp* | *2 tbsp* | *2 tbsp* |

1. Mix together all the ingredients except for the oats and sesame seeds.
2. Form into 4 burgers.
3. Mix together the oats and sesame seeds and roll the burgers in them until well coated.
4. Cook under a moderate grill for 12 minutes each side.

# Oat Vegetable Pie

Serves 4 9 g fibre/280 calories per portion

| | *Metric* | *Imperial* | *American* |
|---|---|---|---|
| ***Swede, diced*** | ***225 g*** | ***8 oz*** | ***2 cups*** |
| ***Carrots, sliced*** | ***225 g*** | ***8 oz*** | ***2 cups*** |
| ***Shelled or frozen peas*** | ***225 g*** | ***8 oz*** | ***2 cups*** |
| ***Parsley, chopped*** | ***2 tbsp*** | ***2 tbsp*** | ***2 tbsp*** |
| ***Vegetable stock*** | ***300 ml*** | ***½ pint*** | ***1¼ cups*** |

**Topping**

| | *Metric* | *Imperial* | *American* |
|---|---|---|---|
| ***Unsalted butter or soft vegetable margarine*** | ***25 g*** | ***1 oz*** | ***2 tbsp*** |
| ***Potato, boiled*** | ***1 large*** | ***1 large*** | ***1 large*** |
| ***Mature English cheddar cheese, grated*** | ***50 g*** | ***2 oz*** | ***½ cup*** |
| ***Oats*** | ***50 g*** | ***2 oz*** | ***½ cup*** |
| ***Wholemeal flour*** | ***100 g*** | ***4 oz*** | ***2 cups*** |
| ***Sea salt*** | | | |
| ***Freshly ground black pepper*** | | | |
| ***Skimmed milk*** | ***1 tsp*** | ***1 tsp*** | ***1 tsp*** |

1. Lightly oil an ovenproof dish and heat the oven to 400°F/200°C (Gas Mark 5).
2. Steam or boil the swede and carrots for 5 minutes. Blanch the fresh peas for 1 minute.
3. Mix the vegetables with the parsley and stock and place in an ovenproof dish. Mash together the fat and potato. Mix the cheese, oats, flour and seasoning and add to the potato.
4. Add enough water to form a soft dough and roll out. Carefully lift it on to the vegetables to form a pastry-style crust.
5. Glaze with milk and bake for 30 minutes.

# TRADITIONAL MOUSSAKA

Serves 6 3 g fibre/360 calories per serving

| | *Metric* | *Imperial* | *American* |
|---|---|---|---|
| ***Onions, finely diced*** | ***2 medium*** | ***2 medium*** | ***2 medium*** |
| ***Lean minced beef*** | ***450 g*** | ***1 lb*** | ***1 lb*** |
| ***Tomato purée*** | ***2 tbsp*** | ***2 tbsp*** | ***2 tbsp*** |
| ***Worcestershire sauce*** | ***1 tsp*** | ***1 tsp*** | ***1 tsp*** |
| ***Oat flour*** | ***2 tbsp*** | ***2 tbsp*** | ***2 tbsp*** |
| ***Water or stock*** | ***150 ml*** | ***$\frac{1}{4}$ pint*** | ***$\frac{2}{3}$ cup*** |
| ***Aubergines*** | ***2*** | ***2*** | ***2*** |
| ***Soft vegetable margarine*** | ***50 g*** | ***2 oz*** | ***$\frac{1}{4}$ cup*** |
| ***Oat flour*** | ***50 g*** | ***2 oz*** | ***$\frac{1}{2}$ cup*** |
| ***Skimmed milk*** | ***300 ml*** | ***$\frac{1}{2}$ pint*** | ***$1\frac{1}{4}$ cups*** |
| ***Mature Cheddar, grated*** | ***50 g*** | ***2 oz*** | ***$\frac{1}{2}$ cup*** |
| ***Free-range egg, lightly beaten*** | ***1*** | ***1*** | ***1*** |
| ***Canned tomatoes*** | ***400 g*** | ***14 oz*** | ***$3\frac{1}{2}$ cups*** |

1. Lightly oil a large oblong ovenproof dish and set oven to 375°F/190°C (Gas Mark 5).
2. Sauté the onion in a little oil and add the beef. Stir and brown well. Add the tomato purée and Worcestershire sauce and sprinkle over the oat flour. Stir well then add the stock and simmer for 15 minutes.
3. Slice the aubergines and blanch in boiling water for 2 minutes, drain.
4. Place the margarine and oat flour in a saucepan and stir over a moderate heat to make a roux. Gradually add the milk, stirring between additions. Remove from heat and stir in the cheese and the egg.
5. Place one third of the aubergines in the bottom of the dish. Place half the meat mixture on top and add the tomatoes. Add another third of the aubergine, the remaining meat mixture and top with the last of the aubergines.
6. Pour over the cheese sauce and bake for 45 minutes until golden brown.

# Chapter 8
# DESSERTS

## LEMON CURD TART

Serves 6 3 g/440 calories per portion

| | *Metric* | *Imperial* | *American* |
|---|---|---|---|
| **Pastry** | | | |
| ***Wholemeal flour*** | ***100 g*** | ***4 oz*** | ***1 cup*** |
| ***Oat flour*** | ***50 g*** | ***2 oz*** | ***½ cup*** |
| ***Ground almonds*** | ***50 g*** | ***2 oz*** | ***½ cup*** |
| ***Soft vegetable margarine*** | ***75 g*** | ***3 oz*** | ***⅓ cup*** |
| ***Grated lemon rind*** | | | |
| ***Water to mix*** | | | |

1. Lightly oil a 22 cm (8 inch) flan ring and preheat oven to 400°F/200°C (Gas Mark 6).
2. Sieve flours into a mixing bowl; stir in almonds. Rub in the fat until the mixture resembles breadcrumbs in consistency.
3. Stir in the lemon rind and enough water to work the mixture into a soft dough. Roll out on a floured surface and line the flan ring. Line the flan dish with greaseproof paper and baking beans and bake blind for 25 minutes.
4. Fill, when cold, with lemon curd.

**Lemon Curd**

| | *Metric* | *Imperial* | *American* |
|---|---|---|---|
| ***Raw cane Muscovado sugar*** | ***100 g*** | ***4 oz*** | ***½ cup*** |
| ***Unsalted butter*** | ***75 g*** | ***3 oz*** | ***⅓ cup*** |
| ***Lemons*** | ***2 large*** | ***2 large*** | ***2 large*** |
| ***Free-range eggs lightly beaten*** | ***2*** | ***2*** | ***2*** |
| ***Additional lemon to garnish*** | | | |

1. Place the sugar, butter, grated rind and juice of the lemons

and eggs in the top of a double boiler or in a basin inside a saucepan of hot water.
2. Stir over a medium heat until the mixture melts and slowly thickens (about 10–15 minutes).
3. Be careful not to boil because this will cause it to curdle.

# FRUIT TART

Serve 4 4 g fibre/255 calories per serving

| | *Metric* | *Imperial* | *American* |
|---|---|---|---|
| ***Wholemeal flour*** | ***100 g*** | ***4 oz*** | ***1 cup*** |
| ***Oat flour*** | ***50 g*** | ***2 oz*** | ***½ cup*** |
| ***Soft vegetable margarine*** | ***50 g*** | ***2 oz*** | ***¼ cup*** |
| ***Cox apples*** | ***4 small*** | ***4 small*** | ***4 small*** |
| ***or other fruit such as plums*** | ***450 g*** | ***1 lb*** | ***1 lb*** |
| ***Clear honey or no-added-sugar jam*** | ***2 tbsp*** | ***2 tbsp*** | ***2 tbsp*** |
| **Topping** | | | |
| ***Soft vegetable margarine*** | ***25 g*** | ***1 oz*** | ***2 tbsp*** |
| ***Clear honey*** | ***1 tbsp*** | ***1 tbsp*** | ***1 tbsp*** |
| ***Oats*** | ***50 g*** | ***2 oz*** | ***½ cup*** |

1. Lightly oil a 22 cm (8 inch) flan ring or dish and heat oven to 375°F/190°C (Gas Mark 5).
2. Sieve the flours into a mixing bowl and rub in the fat until the mixture resembles breadcrumbs in consistency. Bind to a soft dough with a little cold water.
3. Roll out the pastry on a lightly-floured surface and line the prepared flan ring with the pastry. Line it with greaseproof paper weighed down with baking beans and bake blind for 10 minutes. Remove from oven and remove paper and beans.
4. Wash the apples and slice, blanch in boiling water for two minutes and drain.
5. Arrange the slices in the prepared pastry case and brush with the honey or jam.
6. To prepare the topping melt the margarine and honey in a pan, remove from heat and stir in the oats. Sprinkle around the edges of the prepared flan and bake for 20 minutes.

# RHUBARB CRUMBLE

Serves 6 5.5 g fibre/200 calories per portion.

| | *Metric* | *Imperial* | *American* |
|---|---|---|---|
| ***Rhubarb, sliced*** | ***675 g*** | ***1½ lb*** | ***1½ lb*** |
| ***Clear honey*** | ***25 g*** | ***1 oz*** | ***2 tbsp*** |

**Topping**

| | *Metric* | *Imperial* | *American* |
|---|---|---|---|
| ***Wholemeal flour*** | ***75 g*** | ***3 oz*** | ***¾ cup*** |
| ***Rolled oats*** | ***75 g*** | ***3 oz*** | ***¾ cup*** |
| ***Ground cinnamon*** | ***1 tsp*** | ***1 tsp*** | ***1 tsp*** |
| ***Soft vegetable margarine*** | ***75 g*** | ***3 oz*** | ***⅓ cup*** |
| ***Dates, chopped*** | ***50 g*** | ***2 oz*** | ***½ cup*** |

1. Lightly oil an ovenproof dish and heat the oven to 400°F/ 200°C (Gas Mark 6).
2. Wash and slice the rhubarb into 2 cm (1 inch) pieces. Place in a saucepan with 2 tablespoons water and honey. Cover and cook over low heat for 5 minutes. Then transfer the rhubarb to ovenproof dish.
3. Mix together the flour, oats and cinnamon. Rub margarine into the flour mixture and stir in the dates.
4. Bake for 25 minutes.

# Apple Crumble

Serves 6 3 g fibre/235 calories per portion

| | *Metric* | *Imperial* | *American* |
|---|---|---|---|
| ***Wholemeal flour*** | ***75 g*** | ***3 oz*** | ***¾ cup*** |
| ***Rolled oats*** | ***50 g*** | ***2 oz*** | ***½ cup*** |
| ***Sunflower seeds*** | ***1 tbsp*** | ***1 tbsp*** | ***1 tbsp*** |
| ***Desiccated coconut*** | ***1 tbsp*** | ***1 tbsp*** | ***1 tbsp*** |
| ***Soft vegetable margarine*** | ***75 g*** | ***3 oz*** | ***⅓ cup*** |
| ***Cinnamon*** | ***½ tsp*** | ***½ tsp*** | ***½ tsp*** |
| ***Demerara sugar (optional)*** | ***25 g*** | ***1 oz*** | ***2 tbsp*** |
| ***Cooking apples*** | ***675 g*** | ***1½ lb*** | ***1½ lb*** |
| ***Juice of a lemon*** | ***1*** | ***1*** | ***1*** |
| ***Few cloves*** | | | |

1. Lightly oil an ovenproof dish and heat the oven to 400°F/ 200°C (Gas Mark 6).
2. Place flour, oats, seeds and coconut in bowl. Rub in margarine and stir the cinnamon (and sugar if used).
3. Thinly slice the apples and sprinkle with lemon juice as you do so. Place in saucepan and cook gently for two minutes with a few tablespoons of water.
4. Turn into deep ovenproof dish and add the cloves. Cover with topping, smoothing over top, and bake for 20–25 minutes until fruit is tender and topping golden. Serve hot or cold with natural yoghurt.

# Blackberry and Apple Crumble

Serves 4 8.5 g fibre/340 calories per portion

| | *Metric* | *Imperial* | *American* |
|---|---|---|---|
| *Cooking apples* | *450 g* | *1 lb* | *1 lb* |
| *Lemon* | *1* | *1* | *1* |
| *Blackberries* | *225 g* | *8 oz* | *2 cups* |
| *Wholemeal flour* | *50 g* | *2 oz* | *½ cup* |
| *Rolled oats* | *50 g* | *2 oz* | *½ cup* |
| *Unsalted butter or soft vegetable margarine* | *75 g* | *3 oz* | *¾ cup* |
| *Finely chopped dates* | *50 g* | *2 oz* | *½ cup* |
| *Sunflower seeds* | *25 g* | *1 oz* | *2 tbsp* |

1. Lightly oil an ovenproof dish and heat the oven to 400°F/200°C (Gas Mark 6).
2. Core and slice the unpeeled apples and dress with lemon juice to prevent browning. Place in saucepan with the washed blackberries over low heat and cover. Cook for 5 minutes.
3. Turn the fruit into the dish.
4. Mix together the flour and oats and rub the fat into the flour until it resembles breadcrumbs in texture.
5. Stir in the dates and sunflower seeds and place the mixture on top of the fruit.
6. Bake for 25 minutes.

# Crêpe Suzettes

Makes 8/Serves 4 2 g fibre/190 calories per portion

*1 quantity basic oat pancakes, page 48*
*2 large juicy oranges*
*2 large juicy lemons*
*2 tablespoons Cointreau or Grand Marnier*
*2 tablespoons demerara sugar (optional)*

1. Make the crêpes and keep warm.
2. Squeeze the juice from the fruit and mix with the liqueur (and sugar, if using).
3. Fold the crêpes and place in a large frying pan. Pour over the juice and heat through.
4. Flambé, if liked, by adding a little brandy and setting light to the mixture.

# Apple Streudel Ice-cream

Makes 8 scoops     1.5 g fibre/85 calories per scoop

| | Metric | Imperial | American |
|---|---|---|---|
| *Oats* | *50 g* | *2 oz* | *½ cup* |
| *Walnuts, chopped* | *50 g* | *2 oz* | *½ cup* |
| *Cooking apple, peeled, cored and sliced* | *325 g* | *12 oz* | *3 cups* |
| *Juice of ½ a lemon* | | | |
| *Concentrated apple juice* | *2 tbsp* | *2 tbsp* | *2 tbsp* |
| *Natural yoghurt* | *150 ml* | *¼ pint* | *⅔ cup* |
| *Ground cinnamon* | *1 tsp* | *1 tsp* | *1 tsp* |
| *Free-range egg whites* | *2* | *2* | *2* |

1. Grind the oats to flour in a coffee grinder or liquidiser.
2. Place the walnuts, apple, lemon juice and apple juice in a saucepan and cover. Cook over a low heat to make a purée.
3. Remove from heat and beat in the oat flour and cinnamon.
4. Allow to cool and when cold fold in the yoghurt.
5. Whisk the egg whites until stiff and fold into the mixture.
6. Place in a shallow container and freeze. When nearly frozen break up with a fork and mash. Return to the freezer in a container that is deep enough to allow scoops to be taken. Alternatively use an ice-cream making machine.
7. Remove ice-cream from the freezer 20–30 minutes before serving to allow it to soften enough for scoops to be taken.

# Apple Streudel Crêpes

Makes 8 3 g fibre/120 calories per crêpe

| | Metric | Imperial | American |
|---|---|---|---|
| *1 quantity basic oat pancakes, pages 48* | | | |
| *Cooking apple* | *325 g* | *12 oz* | *2 ¾ cups* |
| *Juice of ½ a lemon* | | | |
| *Sultanas* | *50 g* | *2 oz* | *½ cup* |
| *Walnuts, chopped* | *25 g* | *1 oz* | *2 tbsp* |
| *Ground cinnamon* | *½ tsp* | *½ tsp* | *½ tsp* |
| *Water* | *60 ml* | *2 fl. oz* | *¼ cup* |
| *Wholemeal breadcrumbs* | *50 g* | *2 oz* | *½ cup* |

1. Prepare and keep the crêpes warm.

2. Peel the apple and core. Dip slices into lemon juice to prevent browning and place in a stainless steel saucepan with the sultanas, walnuts, cinnamon and water.
3. Cover and simmer for 15 minutes, stirring from time to time and adding more water if necessary.
4. Remove from heat and stir in the breadcrumbs. Spread on to the crêpes and roll up.

# Orange and Apricot Cheesecake

Serves 8 2 g fibre/200 calories per portion

**Pastry**

| | *Metric* | *Imperial* | *American* |
|---|---|---|---|
| *Wholemeal flour* | *100 g* | *4 oz* | *1 cup* |
| *Oat flour* | *50 g* | *2 oz* | *½ cup* |
| *Soft vegetable margarine* | *75 g* | *3 oz* | *⅓ cup* |
| *Water to mix* | | | |

1. Lightly oil a 22 cm (8 inch) flan ring and heat the oven to 400°F/200°C (Gas Mark 6).
2. Sieve the flour into a mixing bowl. Rub in the fat until the mixture resembles breadcrumbs in consistency.
3. Add a little cold water to mix to a soft dough. Roll out the mixture on a floured board. Line the flan ring with pastry and top with greaseproof paper and baking beans. Bake blind for 10 minutes.
4. Lower the temperature of the oven to 375°F/190°C (Gas Mark 5) and prepare filling.

**Filling**

| | *Metric* | *Imperial* | *American* |
|---|---|---|---|
| *Low-fat curd cheese* | *225 g* | *8 oz* | *1 cup* |
| *Milk* | *75 ml* | *2½ fl oz* | *6 tbsp* |
| *Free-range eggs, separated* | *2* | *2* | *2* |
| *Clear honey* | *2 tbsp* | *2 tbsp* | *2 tbsp* |
| *Orange oil essence* | *3 drops* | *3 drops* | *3 drops* |

| | | | |
|---|---|---|---|
| *Grated peel of orange* | *1* | *1* | *1* |
| *Oat flour* | *25 g* | *1 oz* | *2 tbsp* |
| *Free-range egg whites* | *2* | *2* | *2* |
| *Apricot halves* | *8* | *8* | *8* |

1. Place the curd cheese in a mixing bowl and stir in the milk to make a smooth paste.
2. Add the egg yolks, honey, orange oil, orange peel and oat flour. Mix thoroughly until smooth.
3. Whisk the egg whites until firm and stiff. Then fold into the mixture with a metal spoon.
4. Pour into the baked flan case.
5. Arrange the apricot halves on top of the mixture and bake for 20 minutes.

# Currant Cheesecake

Serves 8 2 g fibre/320 calories per portion

Prepare an oat cheesecake base from the Lime Cheesecake, page 74.

**Filling**

| | *Metric* | *Imperial* | *American* |
|---|---|---|---|
| *Cottage cheese* | *225 g* | *8 oz* | *2 cups* |
| *Thin yoghurt or milk* | *150 ml* | *¼ pint* | *⅔ cup* |
| *Soured cream* | *150 ml* | *¼ pint* | *⅔ cup* |
| *Currants* | *50 g* | *2 oz* | *½ cup* |
| *Gelatine* | *12 g* | *½ oz* | *1 tbsp* |
| *Boiling water* | *4 tbsp* | *4 tbsp* | *4 tbsp* |

1. Sieve the cottage cheese into a mixing bowl and stir in the yoghurt or milk and the cream until smooth. Stir in the currants.
2. Sprinkle the gelatine on to the water and stir until dissolved.
3. When on point of setting, stir into the cheesecake mixture and pour on to the prepared base. Leave in the fridge to set.

# LIME CHEESECAKE

Serves 10 1.5 g fibre/270 calories per serving

Decorate this lime cheesecake with strawberries and slices of lime and it becomes a stunning dessert. The oat casing makes an attractive change from the usual biscuit base used in cheesecakes.

**Case**

| | *Metric* | *Imperial* | *American* |
|---|---|---|---|
| ***Soft vegetable margarine*** | ***100 g*** | ***4 oz*** | ***½ cup*** |
| ***Clear honey*** | ***50 g*** | ***2 oz*** | ***¼ cup*** |
| ***Demerara sugar*** | ***50 g*** | ***2 oz*** | ***¼ cup*** |
| ***Rolled oats*** | ***225 g*** | ***8 oz*** | ***2 cups*** |

1. Lightly oil a 22cm (8 inch) spring clip sided cake tin and set the oven to 350°F/180°C (Gas Mark 4).
2. Place the margarine, honey and sugar in a saucepan over a moderate heat and stir until dissolved. Remove from heat.
3. Measure the oats and stir into the mixture. Press into the base and up the sides of the prepared tin and bake for 25 minutes. Remove from oven and cool on a wire tray.

**Filling**

| | *Metric* | *Imperial* | *American* |
|---|---|---|---|
| *Quark* | *225 g* | *8 oz* | *1 cup* |
| *Greek yoghurt, strained* | *225 g* | *8 oz* | *1 cup* |
| *Free-range eggs, separated* | *2* | *2* | *2* |
| *Fructose* | *50 g* | *2 oz* | *$\frac{1}{4}$ cup* |
| *Limes, grated rind and juice* | *2* | *2* | *2* |
| *Gelatine* | *12 g* | *$\frac{1}{2}$ oz* | *1 tbsp* |
| *Boiling water* | *2 tbsp* | *2 tbsp* | *2 tbsp* |

1. Mix the cheese and yoghurt together in a mixing bowl with the egg yolks and the sugar. Stir in the lime rind.
2. Mix the lime juice with the boiling water and sprinkle on the gelatine. Stir until dissolved. When cool and on the point of setting add to the cheese mixture.
3. Whisk the egg whites until stiff. When the cheese mixture is on the point of setting fold in the whites. Pour into the prepared base and leave to set in the fridge.

*Note*: Do not make too far in advance of serving or the base will go soggy.

# DUTCH APPLE CAKE

Serves 6 10 g fibre/450 calories per portion

| | *Metric* | *Imperial* | *American* |
|---|---|---|---|
| *Wholemeal flour* | *175 g* | *6 oz* | *$1\frac{1}{2}$ cups* |
| *Oat flour* | *50 g* | *2 oz* | *$\frac{1}{2}$ cup* |
| *Baking powder* | *1 tsp* | *1 tsp* | *1 tsp* |
| *Unsalted butter or soft vegetable margarine* | *150 g* | *5 oz* | *$1\frac{1}{4}$ cups* |
| *Free-range eggs, separated* | *2* | *2* | *2* |

| | | | |
|---|---|---|---|
| *Lemon* | *1* | *1* | *1* |
| *Cooking apples* | *900 g* | *2 lb* | *2 lb* |
| *Dried apricots, soaked overnight* | *18* | *18* | *18* |
| *Raisins* | *100 g* | *4 oz* | *1 cup* |
| *Cinnamon* | *1 tsp* | *1 tsp* | *1 tsp* |
| *Ground almonds* | *50 g* | *2 oz* | *½ cup* |

1. Lightly oil a 23 cm (9 inch) cake tin and heat oven to 350°F/180°C (Gas Mark 4).
2. Make the pastry by sieving the flour and baking powder into a mixing bowl and rubbing in the fat until the mixture resembles breadcrumbs in consistency.
3. Lightly beat the egg yolks and add to the pastry with a tablespoonful of lemon juice. Bind to a soft dough. Roll out on a lightly-floured board and line the tin. Roll remainder ready for use as lid.
4. Wash and core the apples but do not peel. Slice into a saucepan with a few tablespoons of water and start cooking over a low heat for 5 minutes, stirring to ensure even cooking. Drain. While the apples are cooking, boil the apricots in another pan for about 5 minutes. Drain.
5. Fill the prepared pastry case with layers of apple and apricots and raisins. Mix together the spice and ground almonds and sprinkle on top.
6. Brush the edges of the pie with egg white and place the pastry lid in position. Glaze and bake for 45–50 minutes.

# FRUIT SALAD

| | *Metric* | *Imperial* | *American* |
|---|---|---|---|
| *Apple* | *1* | *1* | *1* |
| *Pear* | *1* | *1* | *1* |
| *Juice of ½ a lemon* | | | |
| *Banana* | *1* | *1* | *1* |
| *Orange juice* | *150 ml* | *¼ pint* | *⅔ cup* |
| *Green grapes* | *175 g* | *6 oz* | *1½ cups* |
| *Dried figs* | *4* | *4* | *4* |
| *Family Favourite Granola, page 37* | *100 g* | *4 oz* | *1 cup* |

Serves 4  2.5 g fibre/200 calories per portion

1. Wash the apple and pear. Quarter, core and dice. Do not peel. Toss in the lemon juice.
2. Peel and slice the banana and add to the apple with the orange juice. Halve the grapes and remove the pips. Slice the figs into thin strips and add to the rest of the salad with the grapes.
3. Just before serving scatter the granola on top of the salad.

# HOT FRUIT COMPÔTÉ

Serves 4  5 g fibre/325 calories per portion

| | *Metric* | *Imperial* | *American* |
|---|---|---|---|
| *Water* | *600 ml* | *1 pint* | *2½ cups* |
| *Clear honey* | *1 tbsp* | *1 tbsp* | *1 tbsp* |
| *Cinnamon stick* | *1* | *1* | *1* |
| *Ground nutmeg* | *2* | *2* | *2* |
| *Cloves* | | | |
| *Lemon, grated rind and juice* | *1* | *1* | *1* |
| *Dried apricots* | *100 g* | *4 oz* | *1 cup* |
| *Dried apple rings* | *100 g* | *4 oz* | *1 cup* |
| *Canned peaches in own (or apple) juice* | *400 g* | *14 oz* | *3½ cups* |
| *Sultanas* | *50 g* | *2 oz* | *½ cup* |
| *Family Favourite Granola, page 37* | *100 g* | *4 oz* | *1 cup* |

1. Place water, honey, spices and lemon in saucepan and bring to boil.
2. Add apricots and apples and simmer, covered for 20 minutes.
3. Add peaches and sultanas and continue cooking for 10 minutes.
4. Serve hot, offering granola in separate dish. Also good with natural yoghurt.

# Black Cherry Fool

Serves 6 2 g fibre/155 calories per serving

| | *Metric* | *Imperial* | *American* |
|---|---|---|---|
| ***Black cherries, stoned*** | ***325 g*** | ***12 oz*** | ***3 cups*** |
| ***Greek yoghurt, strained*** | ***325 g*** | ***12 oz*** | ***1½ cups*** |
| ***Gelatine*** | ***12 g*** | ***½ oz*** | ***1 tbsp*** |
| ***Cold water*** | ***4 tbsp*** | ***4 tbsp*** | ***4 tbsp*** |
| ***Granola, page 37*** | ***6 tbsp*** | ***6 tbsp*** | ***6 tbsp*** |

1. Place the cherries and yoghurt in a liquidiser and blend to a smooth purée.
2. Sprinkle the gelatine on to the water and place in a saucepan over a low heat. Bring to the boil, stirring all the time, until the gelatine has dissolved. This is when the liquid

becomes transparent. Remove from heat and pour into a small basin.

3. Leave to cool and when on point of setting fold or whisk into the purée. Pour into a container and place in the fridge to set.
4. To serve spoon a layer of fool into dessert glasses or sundae dishes and sprinkle with a layer of granola, followed by a second layer of fool. Sprinkle a little layer of granola around the edge of the glass for the top layer. Serve slightly chilled.

# Marzipan Franzipan

Serves 8 3 g fibre/200 calories per portion.

**Pastry**

| | *Metric* | *Imperial* | *American* |
|---|---|---|---|
| ***Wholemeal flour*** | ***50 g*** | ***2 oz*** | ***½ cup*** |
| ***Oat flour*** | ***50 g*** | ***2 oz*** | ***½ cup*** |
| ***Soft vegetable margarine*** | ***50 g*** | ***2 oz*** | ***½ cup*** |
| ***Water to mix*** | | | |

**Filling**

| | *Metric* | *Imperial* | *American* |
|---|---|---|---|
| ***Pears*** | ***3 small*** | ***3 small*** | ***3 small*** |
| ***Raw sugar marzipan*** | ***50 g*** | ***2 oz*** | ***¼ cup*** |
| ***Soft vegetable margarine*** | ***50 g*** | ***2 oz*** | ***¼ cup*** |
| ***Wholemeal flour*** | ***50 g*** | ***2 oz*** | ***½ cup*** |
| ***Baking powder*** | ***1 tsp*** | ***1 tsp*** | ***1 tsp*** |
| ***Free-range egg*** | ***1*** | ***1*** | ***1*** |

1. Preheat the oven to 400°F/200°C (Gas Mark 6) and lightly oil a 22 cm (8 inch) flan ring.
2. To make the pastry, sieve the flours and rub in the margarine until the mixture resembles breadcrumbs in consistency. Bind to a soft dough with water and roll out on a lightly floured surface.

3. Line the flan ring with the pastry. Cover with greaseproof paper and fill with baking beans. Bake blind for 7 minutes. Remove the paper and return to the oven for 3 minutes.
4. Lower oven temperature to 350°F/190°C (Gas Mark 5).
5. To prepare the filling, peel, halve and core the pears, place flat side down in the base of the pastry case. Cream together the marzipan and margarine until soft and light. Sieve together the flour and baking powder. Lightly beat the egg and beat into the marzipan mixture.
6. Fold in the flour, spoon over the pears and return to the oven. Bake for 30 minutes.

# Chantilly Yoghurt Cream

0 g fibre/ 300 calories

This is a deliciously light and low calorie cream which can be used with strawberries or other fruit and gâteaux in place of heavy clotted or double creams. Compare the calorie count of this cream with the 890 calories in 225 g/8 oz/1 cup of double cream.

| | *Metric* | *Imperial* | *American* |
|---|---|---|---|
| ***Greek yoghurt, strained*** | ***225 g*** | ***8 oz*** | ***1 cup*** |
| ***Free-range egg white*** | ***1*** | ***1*** | ***1*** |

1. Scoop the yoghurt into a small mixing bowl.
2. Whisk the egg white until stiff and fold into the yoghurt. Chill before use.

# Piping Chantilly Cream

0 g fibre/ 350 calories

For a stiffer cream that can hold its shape add gelatine to the yoghurt before adding the egg white; this mixture can then be chilled before use.

| | Metric | Imperial | American |
|---|---|---|---|
| ***Greek yoghurt, strained*** | ***225 g*** | ***8 oz*** | ***1 cup*** |
| ***Gelatine*** | ***12 g*** | ***$\frac{1}{2}$ oz*** | ***1 tbsp*** |
| ***Free-range egg white*** | ***1*** | ***1*** | ***1*** |

1. Scoop the yoghurt into a small mixing bowl.
2. Sprinkle the gelatine on to 4 tablespoons of cold water and place in a saucepan. Stir over a low heat until the gelatine has dissolved and the liquid is transparent.
3. When the gelatine has cooled and is on the point of setting stir into the yoghurt.
4. Fold the whisked egg white into the mixture and refrigerate before use.

# BANANA CRÊPE

Makes 8 2 g fibre/ 80 calories per pancake

***1 lemon, grated rind***
***1 quantity of basic oat pancake batter, page 48***
***Pinch of cinnamon***
***2 bananas, thinly sliced***

1. Stir the lemon rind into the basic batter and add a pinch of cinnamon.
2. When the batter has been evenly spread over the base of the crêpe pan add the thin slices of banana to the pancake.
3. Cook on one side for 1–1$\frac{1}{2}$ minutes then slip a fish slice under the pancake and turn to cook briefly on the banana side.

# Strawberry Shortcakes

Makes 5 3 g fibre/350 calories per shortcake

These are a delicious 'sandwich' of strawberries and cream between two short, crispy biscuits. High calorie, double cream is replaced with a light, low-calorie mock cream made from natural yoghurt.

***1 quantity shortbread, page 83***
***1 small punnet fresh strawberries, hulled***
***1 quantity Chantilly piping cream, page 80***

1. Place half the shortbread biscuits on a flat surface and place the strawberries on them, hulled side down.
2. Cut the strawberries to the same height and place the second biscuit on top to check that it is level.
3. Remove the top biscuit and pipe cream into the space between the strawberries just before serving. Replace top biscuit.
4. Offer remaining cream in a separate bowl.

# Shortbread

Makes 10 biscuits 1 g fibre/140 calories per biscuit

| | *Metric* | *Imperial* | *American* |
|---|---|---|---|
| ***Wholemeal flour*** | ***150 g*** | ***5 oz*** | ***1¼ cups*** |
| ***Oat flour*** | ***25 g*** | ***1 oz*** | ***2 tbsp*** |
| ***Fructose or light Muscovado sugar*** | ***40 g*** | ***1½ oz*** | ***2½ tbsp*** |
| ***unsalted butter*** | ***100 g*** | ***4 oz*** | ***½ cup*** |

1. Lightly oil a baking sheet and set the oven to 300°F/150°C (Gas Mark 3).
2. Sieve the flours into a mixing bowl, returning the bran from the sieve to the bowl. Stir in the sugar.
3. Rub the butter into the flour and sugar until the mixture resembles breadcrumbs in consistency. Lightly work to a soft dough. This will be quite crumbly so it will need to be rolled out carefully.
4. Using biscuit cutters cut out the shortbread biscuits and place them on the prepared baking tray by slipping a palette knife beneath them and lifting onto the trays. Bake for 40–45 minutes. Remove from tray and cool on a wire cooling rack. Store in an airtight tin.

**Optional Extras**

For a traditional shortbread round pat and press the shortbread mixture into a 22 cm (8 inch) diameter round and pinch the edges to make decorative indentations. Prick the top with a fork. Shortbread can also be pressed into a square tin and cut into fingers. Or, for a really decorative shortbread use a wooden shortbread mould with a fancy pattern such as a thistle.

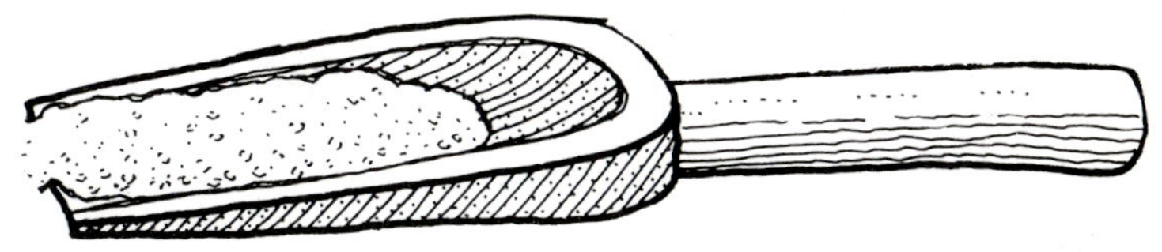

# Chapter 9
# BREAD AND TEATIME TREATS

## TOASTED TEACAKES

Serves 12    3 g fibre/140 calories per portion

| | *Metric* | *Imperial* | *American* |
|---|---|---|---|
| *Wholemeal flour* | *325 g* | *12 oz* | *3 cups* |
| *Oat flour* | *100 g* | *4 oz* | *1 cup* |
| *Mixed spice* | *1 tsp* | *1 tsp* | *1 tsp* |
| *Mixed dried fruit* | *75 g* | *3 oz* | *¾ cup* |
| *Skimmed milk* | *300 ml* | *½ pt* | *1½ cups* |
| *Fresh yeast* | *12 g* | *½ oz* | *1 tbsp* |
| *1, 25 mg vitamin C tablet, crushed* | | | |
| *Free-range egg, beaten with a little milk* | *1* | *1* | *1* |

1. Heat oven to 425°F/220°C (Gas Mark 5).
2. Sieve the flours and spice together into a mixing bowl. Stir the dried fruit into the flour.
3. Warm the milk slightly and crumble in the yeast and the vitamin C. Leave to stand for 5 minutes.
4. Stir the yeast mixture into the flour and add the egg, leaving enough egg to glaze the tops of the teacakes.
5. Mix to soft dough and turn on to a slightly floured surface and knead for 5 minutes. Return to bowl, cover and leave to rest for 10 minutes.
6. Return dough to work surface and knead again. Form into 12 rolls and place on a lightly-oiled baking sheet. Cover and leave to rise until doubled in size (about 10 minutes).
7. Glaze when risen and bake for 20–25 minutes. Before using, cut in half and toast. Then spread with no-added-sugar jam or a little honey.

# DATE AND WALNUT LOAF

Serves 10 1.5 g fibre/260 calories per slice

| | Metric | Imperial | American |
|---|---|---|---|
| *Wholemeal flour* | *175 g* | *6 oz* | *1½ cups* |
| *Oat flour* | *50 g* | *2 oz* | *½ cup* |
| *Baking powder* | *2 tsp* | *2 tsp* | *2 tsp* |
| *Mixed spice* | *1 tsp* | *1 tsp* | *1 tsp* |
| *Ground cinnamon* | *½ tsp* | *½ tsp* | *½ tsp* |
| *Soft vegetable margarine* | *100 g* | *4 oz* | *½ cup* |
| *Muscovado sugar* | *50 g* | *2 oz* | *¼ cup* |
| *Dates, chopped* | *75 g* | *3 oz* | *¾ cup* |
| *Walnuts, chopped* | *75 g* | *3 oz* | *¾ cup* |
| *Free-range eggs* | *2* | *2* | *2* |
| *Skimmed milk* | *2 tbsp* | *2 tbsp* | *2 tbsp* |

1. Heat the oven to 325°F/16°C (Gas Mark 3) and lightly grease a 1 kg (2 lb) loaf tin.
2. Sieve the flour, baking powder, mixed spice and cinnamon into a bowl, adding the bran from the sieve. Rub in the margarine until the mixture resembles fine breadcrumbs.
3. Stir in the sugar and the dates and walnuts.
4. Beat the eggs with the milk and pour on to dry ingredients.
5. Beat well with a wooden spoon to form a soft dough. Place in prepared tin and smooth the top.
6. Place in the centre of the oven and bake for 45–55 minutes.

# SODA BREAD

3 g fibre/100 calories

| | Metric | Imperial | American |
|---|---|---|---|
| *Wholemeal flour* | *450 g* | *1 lb* | *4 cups* |
| *Oat flour* | *225 g* | *8 oz* | *2 cups* |
| *Bicarbonate of soda* | *2 tsp* | *2 tsp* | *2 tsp* |
| *Cultured buttermilk* | *300 ml* | *½ pint* | *1¼ cups* |

1. Lightly oil a baking sheet and heat the oven to 400°F/200°C (Gas Mark 6).
2. Sieve the flours and bicarbonate of soda into a mixing bowl and add the bran from the sieve.
3. Stir in the buttermilk and make a soft dough. Work lightly until it sticks together and mould into a round.
4. Place on a lightly-oiled baking tray and slash the top in the shape of an X. Bake for 40–45 minutes and serve hot.

# GINGERBREAD

Serves 10 1 g fibre/240 calories per slice

| | *Metric* | *Imperial* | *American* |
|---|---|---|---|
| ***Wholemeal flour*** | ***175 g*** | ***6 oz*** | ***1½ cups*** |
| ***Oat flour*** | ***100 g*** | ***4 oz*** | ***1 cup*** |
| ***Bicarbonate of soda*** | ***1 tsp*** | ***1 tsp*** | ***1 tsp*** |
| ***Ground ginger*** | ***2 tsp*** | ***2 tsp*** | ***2 tsp*** |
| ***Ground cinnamon*** | ***1 tsp*** | ***1 tsp*** | ***1 tsp*** |
| ***Soft vegetable margarine*** | ***100 g*** | ***4 oz*** | ***½ cup*** |
| ***Molasses*** | ***3 tbsp*** | ***3 tbsp*** | ***3 tbsp*** |
| ***Clear honey*** | ***2 tbsp*** | ***2 tbsp*** | ***2 tbsp*** |

| | | | |
|---|---|---|---|
| *Skimmed milk with 2 teaspoonfuls lemon juice stirred in* | *150 ml* | *¼ pint* | *⅔ cup* |
| *Free-range eggs* | *2* | *2* | *2* |

1. Lightly oil and line a 22 cm (8 inch) square baking tin and heat the oven to 325°F/160°C (Gas Mark 3).
2. Sieve the flour, bicarbonate of soda and spices into a mixing bowl.
3. Place the margarine, molasses and honey in a saucepan and melt over a low heat. Pour on to the flour and mix well. Add the soured milk and beat in the eggs.
4. Spoon into the tin and bake in the centre of the oven for 50–60 minutes until firm and springy to the touch. Remove from tin, but leave in the paper to cool.

# MADELEINES

Makes 6 3 g fibre/360 calories per madeleine

| | Metric | Imperial | American |
|---|---|---|---|
| *Wholemeal flour* | *75 g* | *3 oz* | *¾ cup* |
| *Oat flour* | *50 g* | *2 oz* | *½ cup* |
| *Soft vegetable margarine* | *100 g* | *4 oz* | *1 cup* |
| *Free-range eggs, lightly beaten* | *2* | *2* | *2* |
| *Clear honey* | *2 tbsp* | *2 tbsp* | *2 tbsp* |
| *Vanilla essence* | *2 drops* | *2 drops* | *2 drops* |
| *Desiccated coconut* | *50 g* | *2 oz* | *½ cup* |
| *No-added-sugar jam, preferably a red one* | | | |

1. Lightly oil 4 dariole moulds and set the oven to 400°F/200°C (Gas Mark 6).
2. Sieve together the flours and place in a mixing bowl.
3. Beat the margarine and honey together until pale and fluffy.
4. Add the vanilla essence and gradually beat in the eggs.
5. Fold in the flour and spoon the mixture into the prepared moulds. Place them on a baking tray and bake in the centre of the oven for 25 minutes.
6. Remove and stand to cool on a wire baking tray before removing the cakes.
7. While still warm brush with warmed jam and roll in desiccated coconut.

# Chelsea Buns

Makes 9 4 g fibre/185 calories per bun

| | Metric | Imperial | American |
|---|---|---|---|
| *Wholemeal flour* | *225 g* | *8 oz* | *2 cups* |
| *Oat flour* | *100 g* | *4 oz* | *1 cup* |
| *Yeast* | *12 g* | *½ oz* | *1 tbsp* |
| *Warm skimmed milk* | *180–240 ml* | *6–8 fl. oz* | *¾–1 cup* |
| *1, 25mg vitamin C tablet, crushed* | | | |
| *Clear honey* | *1 tbsp* | *1 tbsp* | *1 tbsp* |

| | | | |
|---|---|---|---|
| *Free-range egg, lightly beaten* | *1* | *1* | *1* |
| *Mixed dried fruit* | *100 g* | *4 oz* | *1 cup* |
| *Mixed peel* | *50 g* | *2 oz* | *½ cup* |
| *Mixed spice* | *1 tsp* | *1 tsp* | *1 tsp* |
| *Unsalted butter, melted* | *12 g* | *½ oz* | *1 tbsp* |

1. Lightly oil a square baking tin and heat oven to 425°F/220°C (Gas Mark 7).
2. Sieve flour into a mixing bowl.
3. Crumble yeast into the milk and stir in the vitamin C and honey. Pour on to the flour with the beaten egg. Form to a soft dough and then turn on to a lightly floured surface and knead for 5 minutes. Cover the dough and rest it for 15 minutes. Knead lightly and roll out into a 12 x 9 inch rectangle.
4. Mix together the fruit, peel and spice. Brush the dough with the melted butter and scatter with the fruit mixture.
5. Roll the dough into a long 'sausage' and cut into 9 equal pieces. Place them, cut side up, in the tin making a square with 3 rows of 3 pastries. Leave to double in size.
6. Glaze with milk or beaten egg and bake for 20–25 minutes.

# Fruit Cake

Serves 16 3 g fibre/140 calories per slice

| | *Metric* | *Imperial* | *American* |
|---|---|---|---|
| *Soft vegetable margarine* | *175 g* | *6 oz* | *¾ cup* |
| *Raw cane sugar* | *50 g* | *2 oz* | *¼ cup* |
| *Clear honey* | *2 tbsp* | *2 tbsp* | *2 tbsp* |
| *Free-range eggs, beaten* | *3* | *3* | *3* |
| *Wholemeal flour* | *175 g* | *6 oz* | *1½ cups* |
| *Oat flour* | *50 g* | *2 oz* | *½ cup* |
| *Baking powder* | *2 tsp* | *2 tsp* | *2 tsp* |
| *Raisins* | *75 g* | *3 oz* | *¾ cup* |
| *Currants* | *75 g* | *3 oz* | *¾ cup* |
| *Sultanas* | *75 g* | *3 oz* | *¾ cup* |
| *Dried apricots, chopped* | *75 g* | *3 oz* | *¾ cup* |
| *Lemon, juice and rind* | *1* | *1* | *1* |
| *Pecan halves* | *25 g* | *1 oz* | *1 tbsp* |

# FRUIT CAKE continued

1. Lightly oil and line a 22 cm (8 inch) cake tin with greaseproof paper and heat oven to 350°F/180°C (Gas Mark 4).
2. Cream together the margarine, sugar and honey until light and fluffy. Gradually add the eggs with a little flour to prevent the mixture curdling.
3. Sieve the flours and baking powder and mix the dried fruit thoroughly into the flour. Fold the flour and fruit into the creamed mixture and mix in the lemon juice and rind. Spoon into the tin and smooth the top.
4. Place pecan nuts on top and bake in the centre of the oven for 1 hour. Turn down the heat to 325°F/170°C and cover the top of the cake to prevent it over-browning. Bake for a further 30 minutes or until a skewer comes out cleanly.

# BANANA BREAD

Serves 10 2.5 g fibre/225 calories per portion.

| | *Metric* | *Imperial* | *American* |
|---|---|---|---|
| ***Wholemeal flour*** | ***175 g*** | ***6 oz*** | ***1½ cups*** |
| ***Oat flour*** | ***100 g*** | ***4 oz*** | ***1 cup*** |
| ***Baking powder*** | ***1 tsp*** | ***1 tsp*** | ***1 tsp*** |
| ***Mixed spice*** | ***½ tsp*** | ***½ tsp*** | ***½ tsp*** |
| ***Unsalted butter or soft vegetable margarine*** | ***100 g*** | ***4 oz*** | ***½ cup*** |
| ***Barbados sugar*** | ***50 g*** | ***2 oz*** | ***¼ cup*** |
| ***Free-range eggs*** | ***2*** | ***2*** | ***2*** |
| ***Bananas, mashed*** | ***2 large*** | ***2 large*** | ***2 large*** |
| ***Milk*** | ***2 tbsp*** | ***2 tbsp*** | ***2 tbsp*** |

1. Lightly oil a 1 kg (2lb) loaf tin and heat the oven to 375°F/190°C (Gas Mark 5).
2. Sieve the flour, baking powder and spice into a bowl.
3. In another bowl cream the margarine and sugar until light and fluffy. Beat the eggs and gradually add to the margarine mixture. Stir in the banana; fold in the flour.
4. Turn into a lightly oiled loaf tin and bake for 35 minutes.

# SULTANA SCONES

Makes one large round or 12 small scones
1 g fibre/100 calories per scone

| | *Metric* | *Imperial* | *American* |
|---|---|---|---|
| ***Wholemeal flour*** | ***175 g*** | ***6 oz*** | ***1½ cups*** |
| ***Oat flour*** | ***50 g*** | ***2 oz*** | ***½ cup*** |
| ***Baking powder*** | ***1 tsp*** | ***1 tsp*** | ***1 tsp*** |
| ***Unsalted butter or soft vegetable margarine*** | ***50 g*** | ***2 oz*** | ***¼ cup*** |
| ***Sultanas*** | ***50 g*** | ***2 oz*** | ***½ cup*** |
| ***Buttermilk or skimmed milk with few drops of lemon juice added*** | ***150 ml*** | ***¼ pint*** | ***⅔ cup*** |
| ***milk to glaze*** | | | |

1. Lightly oil a baking sheet and heat oven to 425°F/220°C (Gas Mark 7).
2. Sieve the flour and baking powder into a bowl. Rub the butter or margarine into the flour until the mixture resembles breadcrumbs in texture. Stir the sultanas into the mixture.
3. Make a well in the centre of the dry ingredients and gradually add milk, stirring to form a soft dough. Turn the dough on to a lightly floured board and gently knead until dough holds its shape and is pliable. Form into a large round and place on a baking tray. Alternatively roll out to 2 cm (¾ inch) thickness and cut with pastry cutter into 12 scones.
4. Glaze with milk and bake for 15 minutes.

# NO-SUGAR ROCK BUNS

Makes 12 2 g fibre/130 calories per bun

| | *Metric* | *Imperial* | *American* |
|---|---|---|---|
| ***Wholemeal flour*** | ***175 g*** | ***6 oz*** | ***1½ cups*** |
| ***Oat flour*** | ***50 g*** | ***2 oz*** | ***½ cup*** |
| ***Baking powder*** | ***2 tsp*** | ***2 tsp*** | ***2 tsp*** |
| ***Mixed spice or cinnamon*** | ***1 tsp*** | ***1 tsp*** | ***1 tsp*** |
| ***Unsalted butter or soft vegetable margarine*** | ***75 g*** | ***3 oz*** | ***⅓ cup*** |
| ***Sultanas*** | ***50 g*** | ***2 oz*** | ***½ cup*** |
| ***Currants*** | ***50 g*** | ***2 oz*** | ***½ cup*** |
| ***Grated rind of 1 lemon*** | | | |
| ***Milk*** | ***90 ml*** | ***3 fl. oz*** | ***½ cup*** |
| ***Free-range egg*** | ***1*** | ***1*** | ***1*** |

1. Lightly oil a bun tray (12 buns) and heat oven to 400°F/200°C (Gas Mark 6).
2. Sieve the flour, baking powder and spice into mixing bowl. Rub fat into flour until the mixture resembles breadcrumbs in texture. Stir in the sultanas, currants and lemon rind.
3. Stir milk into beaten egg and add to the dry ingredients to make a moist, but not sloppy, consistency. Place a generous teaspoonful of mixture into the bun tins (or individual cake papers) and roughen surface with a fork.
4. Bake for 15–20 minutes.

# OAT BREAD

Makes 1 loaf or 10 rolls 1.5 g fibre/80 calories per roll

| | *Metric* | *Imperial* | *American* |
|---|---|---|---|
| ***Wholemeal flour*** | ***225 g*** | ***8 oz*** | ***2 cups*** |
| ***Oat flour*** | ***100 g*** | ***4 oz*** | ***1 cup*** |
| ***Warm water*** | ***150 ml*** | ***¼ pint*** | ***⅔ cup*** |
| ***Fresh yeast*** | ***12 g*** | ***½ oz*** | ***1 tbsp*** |

| | | | |
|---|---|---|---|
| *1, 25mg vitamin C tablet, crushed* | | | |
| *Corn oil* | *½ tbsp* | *½ tbsp* | *½ tbsp* |
| *Molasses* | *½ tbsp* | *½ tbsp* | *½ tbsp* |
| *Oats* | *12 g* | *½ oz* | *1 tbsp* |

1. Lightly oil a small loaf tin and heat oven to 450°F/230°C (Gas Mark 8).
2. Sieve flour into a mixing bowl.
3. Crumble yeast into water and stir in vitamin C, oil and molasses. Pour on to flour. Mix well and form a dough. Turn on to a lightly-floured surface and knead for 5 minutes.
4. Shape into a long 'sausage' and fold under both ends. Place in prepared tin, cover and leave until doubled in size. Glaze with lightly beaten egg or milk and scatter oats over the top.
5. Bake for 35 minutes until loaf falls easily from tin and sounds hollow when tapped on the base.

# CINNAMON CRUMBLE CAKE

Serves 10 slices 1 g fibre/300 calories per slice

**Cake mixture**

| | *Metric* | *Imperial* | *American* |
|---|---|---|---|
| *Soft vegetable margarine* | *100 g* | *4 oz* | *½ cup* |
| *Clear honey* | *2 tbsp* | *2 tbsp* | *2 tbsp* |
| *Free-range eggs, lightly beaten* | *2* | *2* | *2* |
| *Wholemeal flour, sieved* | *100 g* | *4 oz* | *1 cup* |

**Crumble mixture**

| | *Metric* | *Imperial* | *American* |
|---|---|---|---|
| *Wholemeal flour* | *75 g* | *3 oz* | *¾ cup* |
| *Soft vegetable margarine* | *75 g* | *3 oz* | *⅓ cup* |
| *Rolled oats* | *75 g* | *3 oz* | *¾ cup* |
| *Demerara sugar* | *50 g* | *2 oz* | *¼ cup* |
| *Ground cinnamon* | *1 tsp* | *1 tsp* | *1 tsp* |

1. Lightly oil a 22 cm (8 inch) cake tin and set the oven to 375°F/190°C (Gas Mark 5).
2. To make the sponge cream together the margarine and honey until light and fluffy. Gradually beat in the eggs, adding a little flour if the mixture begins to curdle. Then fold in the flour.
3. To make the crumble sieve the flour into a mixing bowl and rub in the fat until the mixture resembles breadcrumbs in consistency. Stir in the oats, sugar and cinnamon.
4. To assemble place half the sponge in the base of the prepared tin, sprinkle half the crumble mixture on top. Cover with the rest of the sponge, levelling carefully with a spatula or knife, and top with the crumble.
5. Bake for 35–40 minutes. The cake is ready when an inserted skewer comes out clean.
6. Remove from the heat and place the tin to cool on a wire cooling tray. Store in an airtight tin when completely cold.

# Chapter 10
# SNACKS

## SAVOURY SNAPS

Makes 24 biscuits 1 g fibre/65 calories per biscuit

| | *Metric* | *Imperial* | *American* |
|---|---|---|---|
| ***Wholemeal flour*** | ***175 g*** | ***6 oz*** | ***1½ cups*** |
| ***Oat flour*** | ***50 g*** | ***2 oz*** | ***½ cup*** |
| ***Soft vegetable margarine*** | ***50 g*** | ***2 oz*** | ***¼ cup*** |
| ***Sesame seeds*** | ***2 tbsp*** | ***2 tbsp*** | ***2 tbsp*** |
| ***Caraway seeds*** | ***1 tsp*** | ***1 tsp*** | ***1 tsp*** |
| ***Cayenne pepper*** | ***½ tsp*** | ***½ tsp*** | ***½ tsp*** |
| ***Pinch of sea salt*** | | | |
| ***Free-range egg*** | ***1*** | ***1*** | ***1*** |
| ***Little milk to mix*** | | | |

1. Lightly oil two baking sheets and set oven to 375°F/190°C (Gas Mark 5).
2. Place the flour and margarine in a mixing bowl, rub in until the mixture resembles breadcrumbs. Add the remaining ingredients and enough milk to form a soft dough.
3. Roll out to 5 mm (¼ inch) thickness and cut into oblong biscuits.
4. Bake for 15–20 minutes until golden.

## SCOTCH EGGS

Makes 4 4 g fibre/250 calories each

| | *Metric* | *Imperial* | *American* |
|---|---|---|---|
| ***Free-range eggs, hard boiled*** | ***4*** | ***4*** | ***4*** |
| ***Onions, diced*** | ***2*** | ***2*** | ***2*** |
| ***Vegetable oil*** | ***2 tsp*** | ***2 tsp*** | ***2 tsp*** |

| | | | |
|---|---|---|---|
| *Ground hazelnuts* | *50 g* | *2 oz* | *½ cup* |
| *Wholemeal breadcrumbs* | *50 g* | *2 oz* | *½ cup* |
| *Oats* | *50 g* | *2 oz* | *½ cup* |
| *Grated carrots* | *2* | *2* | *2* |
| *Sesame seeds* | *1 tbsp* | *1 tbsp* | *1 tbsp* |
| *Pinch majoram* | | | |
| *Pinch mustard powder* | | | |
| *Tomato ketchup* | *1 tbsp* | *1 tbsp* | *1 tbsp* |
| *Free-range egg to bind mixture, lightly beaten* | *1* | *1* | *1* |

1. Lightly oil a baking tray and heat oven to 375°F/190°C (Gas Mark 5).
2. Shell the hard-boiled eggs.
3. Lightly sauté the onion in the oil until transparent.
4. Mix together in a bowl the hazelnuts, breadcrumbs, oats, carrots, sesame seeds and flavourings. Add the onion and stir in the ketchup and the lightly beaten egg to make a soft mixture.
5. Take a handful of mixture and lightly press one of the hard-boiled eggs into it. Mould the mixture around the egg and place on the baking tray. Repeat with the remaining 3 eggs. Bake for 20-25 minutes.

# Date Slice

Makes 24 1.5 g fibre/125 calories per slice

| | *Metric* | *Imperial* | *American* |
|---|---|---|---|
| *Rolled oats* | *325 g* | *12 oz* | *3 cups* |
| *Unsalted butter or soft vegetable margarine, melted* | *175 g* | *6 oz* | *¾ cup* |
| *Demerara sugar* | *2 tbsp* | *2 tbsp* | *2 tbsp* |
| *Cinnamon* | *1 tsp* | *1 tsp* | *1 tsp* |
| *Pitted dates* | *225 g* | *8 oz* | *2 cups* |
| *Cooking apple* | *1 large* | *1 large* | *1 large* |
| *Water* | *150 ml* | *¼ pint* | *⅔ cup* |

1. Lightly oil a 22 cm (8 inch) square baking tin and heat oven to 375°F/190°C (Gas Mark 5).

# Date Slice continued

2. Place the rolled oats in a mixing bowl and add the melted fat. Mix thoroughly and stir in the sugar and cinnamon.
3. Place half the oat mixture in the base of the tin and press down lightly.
4. Place the dates in a saucepan with the apple which has been sliced but not peeled. Add the water and cook to a pulp.
5. Spread the date purée on top of the oats and top with the rest of the oat mixture. Pat down firmly.
6. Bake for 30 minutes until golden brown.

# CRUNCHY BARS

Makes 12    2 g fibre/220 calories per bar

| | *Metric* | *Imperial* | *American* |
|---|---|---|---|
| ***Soft vegetable margarine*** | ***150 g*** | ***5 oz*** | ***1½ cups*** |
| ***Clear honey*** | ***3 tbsp*** | ***3 tbsp*** | ***3 tbsp*** |
| ***Natural bitter almond essence*** | ***2 drops*** | ***2 drops*** | ***2 drops*** |
| ***Oats*** | ***225 g*** | ***8 oz*** | ***2 cups*** |
| ***Flaked almonds*** | ***25 g*** | ***1 oz*** | ***2 tbsp*** |
| ***Wheatgerm*** | ***50 g*** | ***2 oz*** | ***¼ cup*** |
| ***Sunflower seeds*** | ***25 g*** | ***1 oz*** | ***2 tbsp*** |
| ***Sesame seeds*** | ***25 g*** | ***1 oz*** | ***2 tbsp*** |
| ***Sultanas or raisins*** | ***50 g*** | ***2 oz*** | ***½ cup*** |

1. Lightly oil a 22cm (8 inch) square baking tin and heat oven to 375°F/190°C (Gas Mark 5).
2. Melt margarine and honey in a saucepan. Remove from heat and add almond essence.
3. Place rest of ingredients in a bowl and pour on the margarine. Stir thoroughly and press mixture firmly into the prepared tin.
4. Bake for 20 minutes. Remove from oven and place tin on a baking tray. Cut the mixture into fingers but leave in the tin until completely cold when the fingers should be re-cut and removed.

# CHEESE FLAPJACKS

Makes 14 1 g fibre/125 calories

| | *Metric* | *Imperial* | *American* |
|---|---|---|---|
| ***Rolled oats*** | ***225 g*** | ***8 oz*** | ***2 cups*** |
| ***Soft vegetable margarine*** | ***75 g*** | ***3 oz*** | ***¾ cup*** |
| ***Mature Cheddar cheese, grated*** | ***100 g*** | ***4 oz*** | ***1 cup*** |
| ***Ready-make stoneground mustard*** | ***1 tsp*** | ***1 tsp*** | ***1 tsp*** |
| ***Freshly ground black pepper*** | | | |
| ***Natural yoghurt*** | ***150 ml*** | ***¼ pint*** | ***⅔ cup*** |
| ***Free-range egg, beaten*** | ***1*** | ***1*** | ***1*** |

1. Lightly oil a 22 cm (2 inch) square cake tin and heat oven to 350°F/180°C (Gas Mark 4).
2. Place the oats in a large mixing bowl. Melt the margarine in a saucepan. Stir the cheese and margarine into the oats together with the seasoning, yoghurt and egg.
3. Place in the prepared tin and bake for 25–30 minutes until golden brown.
4. Remove from oven and place tin on a cooling tray. Cut into slices while hot, but do not remove from tin until cold.

# Cheese Crackers

Makes 50  0.5 g fibre/40 calories per cracker

| | Metric | Imperial | American |
|---|---|---|---|
| *Mature cheddar cheese, grated* | *100 g* | *4 oz* | *1 cup* |
| *Parmesan cheese, grated* | *2 tbsp* | *2 tbsp* | *2 tbsp* |
| *Wholemeal flour* | *175 g* | *6 oz* | *1½ cups* |
| *Oat flour* | *75 g* | *3 oz* | *¾ cup* |
| *Soft vegetable margarine* | *75 g* | *3 oz* | *¾ cup* |
| *Dry mustard powder* | *½ tsp* | *½ tsp* | *¼ tsp* |
| *Cold water to mix* | | | |

1. Lightly oil several baking trays and set the oven to 400°F/ 200°C (Gas Mark 6).
2. Stir together in a mixing bowl the cheese and flour.
3. Rub in the margarine and stir in the mustard powder. Mix together with cold water until a soft, but not wet, dough is formed.
4. Lightly flour a flat working surface and roll the dough out thinly. This is probably best done in two batches to make it more manageable.
5. Cut out the biscuits and place them on the trays. Bake for 10 minutes and remove from oven. Place on wire cooling trays to cool and crispen.

# Cheese Scones

Makes 6  2 g fibre/115 calories per scone

| | Metric | Imperial | American |
|---|---|---|---|
| *Oat flour* | *25 g* | *1 oz* | *2 tbsp* |
| *Wholemeal flour* | *75 g* | *3 oz* | *¾ cup* |
| *Baking powder* | *1 tsp* | *1 tsp* | *1 tsp* |
| *Dry mustard powder* | *½ tsp* | *½ tsp* | *½ tsp* |
| *Paprika* | *¼ tsp* | *¼ tsp* | *¼ tsp* |
| *Soft vegetable margarine or unsalted butter* | *25 g* | *1 oz* | *1 tbsp* |

| | | | |
|---|---|---|---|
| *Cheddar cheese, grated* | *50 g* | *2 oz* | *2 tbsp* |
| *Skimmed milk or beaten egg for glazing* | | | |
| *Oats* | *2 tbsp* | *2 tbsp* | *2 tbsp* |
| *Paprika* | *½ tsp* | *½ tsp* | *½ tsp* |

1. Lightly oil a baking tray and heat the oven to 425°F/220°C (Gas Mark 7).
2. Sieve the flour into a mixing bowl with the baking powder, mustard powder and paprika.
3. Rub the fat into flour until the mixture resembles breadcrumbs in consistency. Stir in the grated cheese.
4. Gradually blend in the milk, using a fork, until the mixture is soft and pliable. Place the dough on a lightly floured surface and roll out to about 2 cm (¾ inch) thickness.

5. Cut the scones with a plain or crinkle-edged cutter and glaze the tops with milk or beaten egg.
6. Scatter over the oats and paprika. Place on baking tray and bake for 10–12 minutes.

# Malt Loaf

Serves 10 slices 2.5 g fibre/160 calories per slice

| | *Metric* | *Imperial* | *American* |
|---|---|---|---|
| ***Soft vegetable margarine*** | ***25 g*** | ***1 oz*** | ***2 tbsp*** |
| ***Molasses*** | ***2 tbsp*** | ***2 tbsp*** | ***2 tbsp*** |
| ***Sunwheel malt barley syrup, or malt extract*** | ***3 tbsp*** | ***3 tbsp*** | ***3 tbsp*** |

| | | | |
|---|---|---|---|
| *Skimmed milk* | *150 ml* | *¼ pint* | *⅔ cup* |
| *Wholemeal flour* | *175 g* | *6 oz* | *1½ cups* |
| *Oat flour* | *75 g* | *3 oz* | *¾ cup* |
| *Baking powder* | *2 tsp* | *2 tsp* | *2 tsp* |
| *Seedless raisins* | *100 g* | *4 oz* | *1 cup* |

1. Lightly oil a large loaf tin and set the oven to 350°F/180°C (Gas Mark 4).
2. Place the margarine, molasses, malt barley syrup and milk in a saucepan and melt together over a low heat.
3. Sieve the flours and baking powder into a mixing bowl and stir in the raisins.
4. Pour the liquid into the bowl and mix thoroughly.
5. Spoon into the loaf tin and level the top. Bake in the centre of the oven for 50 minutes or until an inserted skewer comes out clean.
6. Transfer to a wire baking tray to cool and remove from tin when cold.

# SESAME FLAPJACKS

Makes 16 1.5 g fibre/160 calories per flapjack

| | *Metric* | *Imperial* | *American* |
|---|---|---|---|
| *Sesame seeds, toasted* | *50 g* | *2 oz* | *½ cup* |
| *Soft vegetable margarine* | *150 g* | *5 oz* | *1½ cups* |
| *Clear honey* | *4 tbsp* | *4 tbsp* | *4 tbsp* |
| *Demerara sugar* | *50 g* | *2 oz* | *¼ cup* |
| *Desiccated coconut* | *50 g* | *2 oz* | *½ cup* |
| *Oats* | *175 g* | *6 oz* | *1½ cups* |

1. Lightly oil a swiss roll tin and heat oven to 325°F/170°C (Gas Mark 3).
2. Place margarine, honey and sugar in a saucepan and heat until margarine melts. Stir in the coconut, oats and sesame seeds, mix well and spread into prepared tin, levelling the top.
3. Bake in the centre of the oven for 30 minutes until golden.
4. Remove from heat and cut through into slices, but leave in tin until completely cold and firm before removing.

# Honey Flapjacks

Makes 10 1.5 g fibre/180 calories per flapjack

| | *Metric* | *Imperial* | *American* |
|---|---|---|---|
| ***Soft vegetable margarine*** | ***75 g*** | ***3 oz*** | ***⅓ cup*** |
| ***Clear honey*** | ***150 g*** | ***5 oz*** | ***1¼ cups*** |
| ***Porridge oats*** | ***225 g*** | ***8 oz*** | ***2 cups*** |

1. Lightly oil a 22 cm (8 inch) square cake tin and heat oven to 350°F/180°C (Gas Mark 4).
2. Place margarine and honey in large saucepan and heat gently until the margarine has melted. Stir in the oats.
3. Place mixture in tin, pressing down well and smoothing top. Bake in oven for 20–25 minutes until golden brown. Place on a wire tray and mark into fingers.
4. Leave in tin until quite cold before removing and storing in an airtight tin.

# Cheese and Onion Pancakes

Makes 12 1 g fibre/65 calories per pancake

| | *Metric* | *Imperial* | *American* |
|---|---|---|---|
| ***1 quantity of basic oat pancake batter, page 48*** | | | |
| ***Potato*** | ***1 medium*** | ***1 medium*** | ***1 medium*** |
| ***Onion*** | ***1 medium*** | ***1 medium*** | ***1 medium*** |
| ***Red Leicester cheese*** | ***50 g*** | ***2 oz*** | ***½ cup*** |

1. Make the batter in the usual way and grate into it a scrubbed potato, diced onion and the grated cheese.
2. Heat oil in frying pan and when hot add tablespoonfuls of the batter mixture. Cook for 1–2 minutes on either side. Serve hot.

# Yoghurt Treat

Serves 1 3.5 g fibre/210 calories per portion.

| | *Metric* | *Imperial* | *American* |
|---|---|---|---|
| ***Dates, fresh*** | ***4*** | ***4*** | ***4*** |
| ***Natural yoghurt*** | ***150 ml*** | ***¼ pint*** | ***⅔ cup*** |
| ***Clear honey*** | ***1 tsp*** | ***1 tsp*** | ***1 tsp*** |
| ***Family Favourite Granola, page 37*** | ***2 tsp*** | ***2 tsp*** | ***2 tsp*** |

1. Wash the dates, stone and cut into 4. Mix the dates and yoghurt and place in a serving glass. Drizzle over the honey.
2. Top with granola and serve at once.

# Oatcake Fingers

Makes 24 0.5 g fibre/55 calories per finger

| | *Metric* | *Imperial* | *American* |
|---|---|---|---|
| ***Medium oatmeal*** | ***200 g*** | ***7 oz*** | ***1¾ cups*** |
| ***Oat flour*** | ***75 g*** | ***3 oz*** | ***¾ cup*** |
| ***Baking powder*** | ***1 tsp*** | ***1 tsp*** | ***1 tsp*** |
| ***Soft vegetable margarine*** | ***50 g*** | ***2 oz*** | ***⅓ cup*** |
| ***Boiling water to mix*** | | | |

1. Lightly oil two baking trays and heat oven to 375°F/190°C (Gas Mark 5).
2. Mix oatmeal, oat flour and baking powder in a mixing bowl. Melt margarine in a saucepan and stir into dry ingredients. Gradually add enough boiling water to make a dough, being careful not to add too much. Turn on to a lightly floured surface and knead until dough is firm enough to roll out.
3. Roll dough into a rectangle. Trim edges and cut into fingers. Slip palette knife under fingers and carefully lift on to the baking trays. Bake for 10-15 minutes.

# HERB SCONES

Make 6 1.5 g fibre/100 calories per scone.

| | Metric | Imperial | American |
|---|---|---|---|
| *Wholemeal flour* | *75 g* | *3 oz* | *¾ cup* |
| *Oat flour* | *25 g* | *1 oz* | *2 tbsp* |
| *Baking powder* | *1 tsp* | *1 tsp* | *1 tsp* |
| *Soft vegetable margarine* | *40 g* | *1½ oz* | *⅓ cup* |
| *Freshly chopped sage, thyme, parsley, or other herb of choice* | *2 tsp* | *2 tsp* | *2 tsp* |
| *Skimmed milk* | *70 ml* | *2½ fl oz* | *⅓ cup* |

1. Lightly oil a baking tray and heat oven to 425°F/220°C (Gas Mark 7).
2. Sieve the flour into a mixing bowl with the baking powder. Rub the margarine into the flour until mixture resembles breadcrumbs in texture.
3. Stir in the herbs and the milk and knead lightly until firm enough to roll out on a floured surface. Roll dough to about 1 cm (½ inch) thickness and cut out the scones.
4. Place on the baking tray and brush the top of the scones with milk. Bake for 10 minutes. Serve hot or cold.

# BLINIS

Makes 24    1 g fibre/55 calories per blini

Blinis are pancakes traditionally made of buckwheat flour and yeast and served with caviar and soured cream. They are thicker in texture than crêpes and pancakes and tend to look more like drop scones. They are a useful supper dish (without the caviar!) and can be served with savoury sauces, grated vegetables or sprinkled with a little cheese.

| | *Metric* | *Imperial* | *American* |
|---|---|---|---|
| ***Wholemeal flour*** | ***250 g*** | ***10 oz*** | ***2½ cups*** |
| ***Oat flour*** | ***50 g*** | ***2 oz*** | ***½ cup*** |
| ***Fresh yeast*** | ***12 g*** | ***½ oz*** | ***1 tbsp*** |
| ***Lukewarm water*** | ***300 ml*** | ***½ pint*** | ***1½ cups*** |
| ***Unsalted butter, melted*** | ***25 g*** | ***1 oz*** | ***2 tbsp*** |
| ***Free-range eggs, separated*** | ***2*** | ***2*** | ***2*** |
| ***Lukewarm skimmed milk*** | ***300 ml*** | ***½ pint*** | ***1½ cups*** |

1. Sieve the flours and place half in a mixing bowl.
2. Crumble the yeast into the water and stir into the flour. Mix well, cover and leave in a warm place for 30 minutes to rise and double in volume.
3. When risen add the remaining flour, butter and beaten egg yolks. Beat the mixture until smooth. Stir in the warmed milk and beat well. Cover and leave to rise as before.
4. Heat oil in a frying pan and put 2 tablespoonfuls of batter in the pan to make 2 blinis of about 10 cm (4 inches) wide.
5. Whisk the egg whites until stiff and fold into the batter.
6. Cook pancakes for 1 minute on each side.

# Chapter 11
# BISCUITS

## CHRISTENING BISCUITS

Makes 36 biscuits    1 g fibre/65 calories per biscuit

| | *Metric* | *Imperial* | *American* |
|---|---|---|---|
| ***Wholemeal flour*** | ***225 g*** | ***8 oz*** | ***2 cups*** |
| ***Oat flour*** | ***100 g*** | ***4 oz*** | ***1 cup*** |
| ***Mixed spice*** | ***1 tsp*** | ***1 tsp*** | ***1 tsp*** |
| ***Free-range eggs*** | ***2*** | ***2*** | ***2*** |
| ***Soft vegetable margarine*** | ***100 g*** | ***4 oz*** | ***½ cup*** |
| ***Demerara sugar*** | ***50 g*** | ***2 oz*** | ***¼ cup*** |
| ***Currants*** | ***100 g*** | ***4 oz*** | ***1 cup*** |
| ***Skimmed milk*** | ***150 ml*** | ***¼ pint*** | ***⅔ cup*** |

1. Lightly oil 2 baking trays and heat oven to 350°F/180°C (Gas Mark 4).
2. Sieve together the flour and spice.
3. Separate the eggs, whisking the whites until stiff but not dry.
4. Cream together the fat and sugar until pale and fluffy in texture. Add the egg yolks and stir in the currants.
5. Stir in the flour and milk. Fold in the whisked egg whites.
6. Roll out the biscuit dough on a lightly floured surface and cut out using a medium-sized fancy edged cutter.
7. Place biscuits on the trays and glaze with egg white. Bake for 20 minutes.

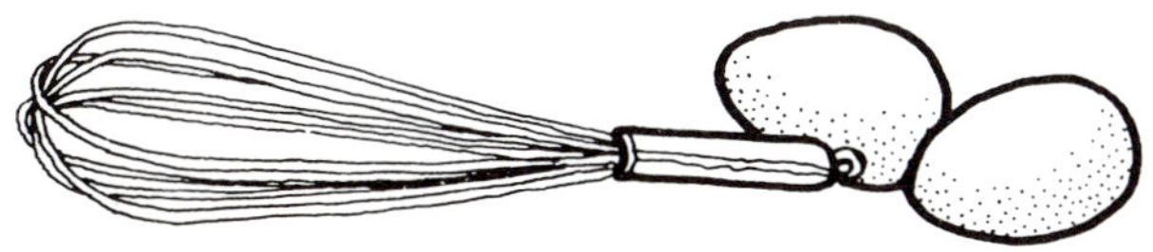

# Oatmeal Biscuits

Makes 30 biscuits 1 g fibre/52 calories per biscuit.

| | Metric | Imperial | American |
|---|---|---|---|
| *Oatmeal* | *200 g* | *7 oz* | *1¾ cups* |
| *Wholemeal flour* | *75 g* | *3 oz* | *¾ cup* |
| *Oat flour* | *50 g* | *2 oz* | *½ cup* |
| *Demerara sugar* | *50 g* | *2 oz* | *½ cup* |
| *Baking powder* | *1 tsp* | *1 tsp* | *1 tsp* |
| *Soft vegetable margarine* | *100 g* | *4 oz* | *½ cup* |
| *Free-range egg* | *1* | *1* | *1* |
| *Skimmed milk* | *2 tbsp* | *2 tbsp* | *2 tbsp* |

1. Lightly oil 2 baking trays and heat oven to 375°F/190°C (Gas Mark 5).
2. Mix together the oatmeal, flour, sugar and baking powder. Rub in the fat to make breadcrumb consistency.
3. Make well in centre of ingredients and add egg and milk. Mix to stiff dough and roll out thinly on a lightly floured surface.
4. Cut biscuits with cutter and place on baking tray. Glaze with milk or beaten egg.
5. Bake for 15 minutes until golden brown.

# Coffee Crunch

Makes 15 1 g fibre/90 calories per biscuit

| | Metric | Imperial | American |
|---|---|---|---|
| *Wholemeal flour* | *100 g* | *4 oz* | *1 cup* |
| *Oat flour* | *50 g* | *2 oz* | *½ cup* |
| *Baking powder* | *1 tsp* | *1 tsp* | *1 tsp* |
| *Soft vegetable margarine* | *50 g* | *2 oz* | *¼ cup* |
| *Fructose* | *50 g* | *2 oz* | *½ cup* |
| *Finely chopped walnuts* | *50 g* | *2 oz* | *½ cup* |
| *Free-range egg, beaten* | *1* | *1* | *1* |
| *2 teaspoons decaffeinated coffee granules dissolved in 4 teaspoons boiling water* | | | |

1. Lightly oil a baking tray and heat oven to 350°F/180°C (Gas Mark 4).
2. Sieve the flour and baking powder into a mixing bowl.
3. Rub in the fat until the mixture resembles breadcrumbs.
4. Stir in the fructose and walnuts.
5. Beat in the egg and add the coffee mixture.
6. Roll out the dough and using plain or crinkle-edged cutters cut out the biscuits. Place them on a baking tray and bake for 20 minutes.

# Honey Oaties

Makes 30 0.5 g fibre/55 calories per biscuit.

| | *Metric* | *Imperial* | *American* |
|---|---|---|---|
| ***Soft vegetable margarine*** | ***100 g*** | ***4 oz*** | ***½ cup*** |
| ***Clear honey*** | ***2 tsp*** | ***2 tsp*** | ***2 tsp*** |
| ***Oat flour*** | ***100 g*** | ***4 oz*** | ***1 cup*** |
| ***Rolled oats*** | ***150 g*** | ***5 oz*** | ***1¼ cups*** |

1. Lightly oil 2 baking trays and heat oven to 375°F/190°C (Gas Mark 5).
2. Melt the margarine and honey together in a saucepan over a low heat.
3. Stir together in a bowl the oat flour and rolled oats.
4. Pour in the honey mixture and mix to a paste.
5. Place large teaspoonfuls of mixture on baking trays. Flatten with the back of a fork and bake for 12 minutes or until golden brown.
6. Remove from oven and slip a palette knife under the biscuits. Transfer them to a wire cooling rack to crispen. Store in an airtight tin.

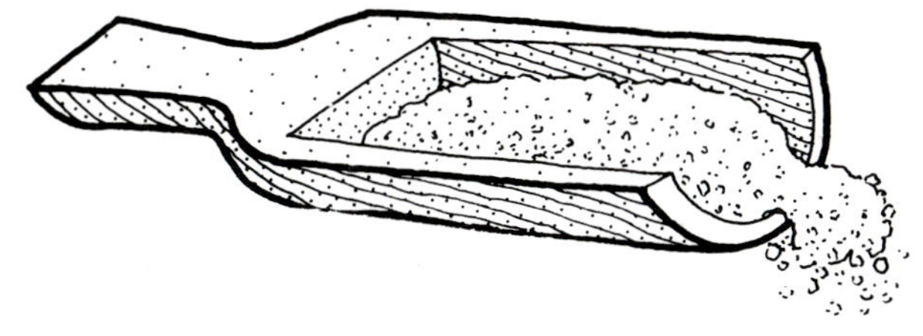

# Piped Oaties

Makes 20  0.5 g fibre/50 calories per biscuit

| | Metric | Imperial | American |
|---|---|---|---|
| *Soft vegetable margarine* | *50 g* | *2 oz* | *¼ cup* |
| *Clear honey* | *2 tbsp* | *2 tbsp* | *2 tbsp* |
| *Wholemeal flour* | *75 g* | *3 oz* | *¾ cup* |
| *Oat flour* | *50 g* | *2 oz* | *½ cup* |
| *Baking powder* | *1 tsp* | *1 tsp* | *1 tsp* |
| *Free-range egg white, whisked* | *1* | *1* | *1* |

1. Lightly oil 2 baking trays and heat oven to 425°F/220°C (Gas Mark 7).
2. Cream margarine and honey until pale and fluffy. Sieve together the flours and baking powder. Beat the flour into the margarine and fold in the egg white.
3. Spoon the mixture into a piping bag with a 1 cm (½ inch) nozzle and pipe small blobs on to the baking tray.
4. Bake for 15 minutes and cool on a wire tray then store in an airtight tin.

# Crunchy Carob Biscuits

Makes 20  0.5 g fibre/63 calories per biscuit

| | Metric | Imperial | American |
|---|---|---|---|
| *Soft vegetable margarine* | *100 g* | *4 oz* | *½ cup* |
| *Raw cane sugar* | *50 g* | *2 oz* | *¼ cup* |
| *Wholemeal flour* | *50 g* | *2 oz* | *½ cup* |
| *Oat flour* | *50 g* | *2 oz* | *½ cup* |
| *Carob powder* | *25 g* | *1 oz* | *2 tbsp* |
| *2 teaspoons decaffeinated coffee granules dissolved in 2 teaspoons hot water* | | | |

1. Lightly oil baking tray and heat oven to 375°F/190°C (Gas Mark 5).
2. Cream together the margarine and sugar until fluffy.
3. Sieve together the flour and carob powder and add to the creamed mixture with the coffee to make a soft paste.
4. With floured hands, lightly roll out walnut-sized pieces of paste and place on the baking tray. Press down to flatten using the back of a fork dipped in hot water.
5. Bake for 12 minutes. Allow to cool on a wire tray and store in an airtight tin.

# Hazelnut And Carob Squares

Makes 14 1 g fibre/ 80 calories per biscuit

| | *Metric* | *Imperial* | *American* |
|---|---|---|---|
| ***Soft vegetable margarine*** | ***50 g*** | ***2 oz*** | ***¼ cup*** |
| ***Muscovado sugar*** | ***50 g*** | ***2 oz*** | ***¼ cup*** |
| ***Free-range egg, beaten*** | ***1*** | ***1*** | ***1*** |
| ***Wholemeal flour*** | ***50 g*** | ***2 oz*** | ***½ cup*** |
| ***Oat flour*** | ***50 g*** | ***2 oz*** | ***½ cup*** |
| ***Carob powder*** | ***2 tsp*** | ***2 tsp*** | ***2 tsp*** |
| ***Baking powder*** | ***1 tsp*** | ***1 tsp*** | ***1 tsp*** |
| ***Toasted hazelnuts, chopped*** | ***50 g*** | ***2 oz*** | ***½ cup*** |

1. Lightly oil a baking tray and heat oven to 350°F/180°C (Gas Mark 4).
2. Cream together the fat and sugar until pale and fluffy. Beat in the egg and sieve together flour, carob powder and baking powder.
3. Fold flour into the mixture and stir in the nuts to form a soft dough.
4. Roll out on a lightly floured surface to 5 mm (⅛ inch) thickness and cut into squares with a biscuit (or ravioli) cutter.
5. Place on baking tray and bake for 20 minutes.

# GINGER BISCUITS

Makes 16 0.5 g fibre/60 calories per biscuit

| | *Metric* | *Imperial* | *American* |
|---|---|---|---|
| ***Wholemeal flour*** | ***50g*** | ***2 oz*** | ***½ cup*** |
| ***Oat flour*** | ***50 g*** | ***2 oz*** | ***½ cup*** |
| ***Baking powder*** | ***1 tsp*** | ***1 tsp*** | ***1 tsp*** |
| ***Ground ginger*** | ***25 g*** | ***1 oz*** | ***2 tbsp*** |
| ***Muscovado sugar*** | ***1 tsp*** | ***1 tsp*** | ***1 tsp*** |
| ***Soft vegetable margarine*** | ***50 g*** | ***2 oz*** | ***¼ cup*** |
| ***Clear honey*** | ***2 tbsp*** | ***2 tbsp*** | ***2 tbsp*** |

1. Lightly oil a baking tray and heat oven to 375°F/190°C (Gas Mark 5).
2. Sieve the flour, baking powder and ginger into a bowl.
3. Place the sugar, margarine and honey in a saucepan and melt over a gentle heat. Stir in the flour.
4. Place generous teaspoonfuls on the baking tray. Press tops lightly with a hot fork to make a pattern. Bake for 15 minutes.

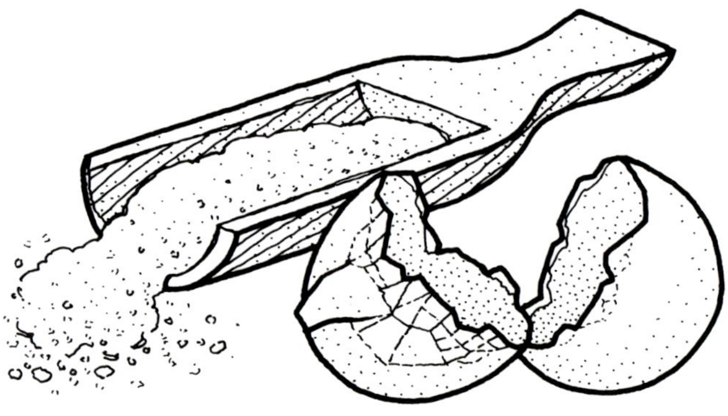

# *Chapter 12*
# CHRISTMAS AND EASTER

## CHRISTMAS PUDDINGS

Makes 2 $1\frac{1}{2}$ pint (900 ml) puddings - each pudding serves 8 1.5 g fibre/155 calories per serving

| | Metric | Imperial | American |
|---|---|---|---|
| ***Raw cane sugar*** | *75 g* | *3 oz* | *⅓ cup* |
| ***Soft vegetable margarine*** | *75 g* | *3 oz* | *⅓ cup* |
| ***Free-range eggs*** | *6* | *6* | *6* |
| ***Wholemeal flour*** | *75 g* | *3 oz* | *¾ cup* |
| ***Oat flour*** | *50 g* | *2 oz* | *½ cup* |
| ***Ginger, mixed spice, and nutmeg*** | *1 tsp each* | *1 tsp each* | *1 tsp each* |
| ***Unblanched almonds, finely chopped*** | *25 g* | *1 oz* | *¼ cup* |
| ***Sultanas*** | *75 g* | *3 oz* | *¾ cup* |
| ***Currants*** | *75 g* | *3 oz* | *¾ cup* |
| ***Raisins*** | *50 g* | *2 oz* | *½ cup* |
| ***Wholemeal breadcrumbs*** | *50 g* | *2 oz* | *½ cup* |
| ***Lemons, juice and rind*** | *2* | *2* | *2* |
| ***Brandy, rum or milk*** | *2 tbsp* | *2 tbsp* | *2 tbsp* |

1. Lightly butter 2, 900 ml (1½ pint) pudding basins and prepare a double layer of greaseproof paper to tie over the top, covered with either a pudding cloth or layer of aluminium foil. Have ready a large saucepan in which the puddings can stand to be steamed, or a pressure cooker.
2. Beat together the sugar and margarine until light and fluffy in consistency. Lightly beat the eggs.
3. Sieve the flour and spices and gradually add the egg and flour, alternately.
4. Fold in the nuts and dried fruit. Add the breadcrumbs and stir well.
5. Finally mix in the lemon juice and rind and the brandy, rum or milk.
6. Divide the mixture between the basins and cover. Steam for 3 hours.
7. Remove from the heat and take off the wet covering. Allow to become completely cold before recovering with clean paper and cloths. Alternatively use pudding basins with fitted lids.
8. Steam for 1 hour before serving. If using pressure cooker steam at highest pressure for 1 hour before serving.

# Christmas Cake

Serves 20 4 g fibre/200 calories per serving

| | *Metric* | *Imperial* | *American* |
|---|---|---|---|
| ***Soft vegetable margarine*** | ***225 g*** | ***8 oz*** | ***1 cup*** |
| ***Raw cane sugar*** | ***100 g*** | ***4 oz*** | ***½ cup*** |
| ***Free-range eggs*** | ***4*** | ***4*** | ***4*** |
| ***Wholemeal flour*** | ***175 g*** | ***6 oz*** | ***1½ cups*** |
| ***Oat flour*** | ***50 g*** | ***2 oz*** | ***½ cup*** |
| ***Mixed spice*** | ***1 tsp*** | ***1 tsp*** | ***1 tsp*** |
| ***Ground almonds*** | ***50 g*** | ***2 oz*** | ***½ cup*** |
| ***Currants*** | ***175 g*** | ***6 oz*** | ***1½ cups*** |
| ***Sultanas*** | ***175 g*** | ***6 oz*** | ***1½ cups*** |
| ***Raisins*** | ***175 g*** | ***6 oz*** | ***1½ cups*** |
| ***Mixed peel*** | ***50 g*** | ***2 oz*** | ***½ cup*** |
| ***Unblanched almonds, finely chopped*** | ***50 g*** | ***2 oz*** | ***½ cup*** |
| ***Stem ginger, finely diced*** | ***50 g*** | ***2 oz*** | ***½ cup*** |
| ***Brandy or rum*** | ***4 tbsp*** | ***4 tbsp*** | ***4 tbsp*** |
| ***Lemon, grated rind and juice*** | ***1*** | ***1*** | ***1*** |

1. Lightly oil and line with a double layer of greaseproof paper 22 cm (8 inch) cake tin and set the oven to 325°F/160°C (Gas Mark 3).
2. Cream together the margarine and sugar until light and fluffy.
3. Lightly beat the eggs and sieve the flours and spices. Gradually add them, alternately, to the margarine mixture.
4. Stir in the ground almonds and the fruit, nuts and ginger and stir in the brandy, lemon rind and juice.
5. Spoon the mixture into the prepared tin and level the top leaving a slight indentation in the centre to allow for the cake to rise.
6. Bake for 3 hours. Cover, if necessary, with paper for the last 45 minutes of baking to prevent over-browning.

**To decorate**

Instead of the usual marzipan and royal icing why not opt for an attractive topping of nuts and dried fruit? This is not only a lot healthier, being lower in sugar and fats and higher

in fibre, but it also looks good and will be a pleasant change for visitors. For a finishing touch use a Christmas cake frill or a red ribbon around the cake.

# MINCE PIES

Makes 12 1 g fibre/125 calories per pie

| | Metric | Imperial | American |
|---|---|---|---|
| ***Wholemeal flour*** | ***100 g*** | ***4 oz*** | ***1 cup*** |
| ***Oat flour*** | ***50 g*** | ***2 oz*** | ***½ cup*** |
| ***Soft vegetable margarine*** | ***75 g*** | ***3 oz*** | ***¾ cup*** |
| ***Cold water to mix*** | | | |
| ***Mincemeat*** | ***75 g*** | ***3 oz*** | ***¾ cup*** |
| ***Egg wash or milk*** | | | |

1. Lightly oil a bun tray (for 12 buns) and set oven to 400°F/ 200°C (Gas Mark 6).
2. Sieve flour into a mixing bowl and rub in fat until mixture resembles breadcrumbs in consistency.
3. Bind with enough water to make a soft, but not wet dough, and turn on to a lightly floured surface.
4. Roll out the pastry and using pastry cutters cut 12 bases and 12 slightly smaller tops for the pies.
5. Place the bases in the lightly prepared bun tin and place a teaspoonful of mincemeat in each base. Brush the edges with water and place a top in position, pressing down to seal the sides.
6. Glaze the tops of the pies with egg wash and bake for 20–25 minutes.

# Hot Cross Buns

Makes 16    3 g fibre/125 calories per bun

| | Metric | Imperial | American |
|---|---|---|---|
| *Wholemeal flour* | *325 g* | *12 oz* | *3 cups* |
| *Oat flour* | *75 g* | *3 oz* | *¾ cup* |
| *Fresh yeast* | *25 g* | *1 oz* | *2 tbsp* |
| *Skimmed milk* | *180 ml* | *6 fl oz* | *1 cup* |
| *Free-range eggs* | *2* | *2* | *2* |
| *Sultanas* | *75 g* | *3 oz* | *¾ cup* |
| *Currants* | *75 g* | *3 oz* | *¾ cup* |
| *Mixed peel* | *50 g* | *2 oz* | *¾ cup* |
| *Ground cinnamon* | *1 tsp* | *1 tsp* | *1 tsp* |
| *Mixed spice* | *1 tsp* | *1 tsp* | *1 tsp* |

**To decorate**

| | Metric | Imperial | American |
|---|---|---|---|
| *Pastry trimmings* | | | |
| *Oats* | | | |
| *Beaten egg or milk to glaze* | | | |
| *Clear honey* | *1 tbsp* | *1 tbsp* | *1 tbsp* |

1. Sieve together the flour and spice and place in a mixing bowl. Heat the milk to blood heat and crumble in the yeast. Stir well.
2. Leave to stand for 10 minutes then pour into the flour. Add the beaten eggs and beat to a dough. Turn on to a lightly floured surface and knead for 5 minutes. Return to the bowl, cover and leave to rest for 10 minutes.
3. Lightly oil 2 baking trays and set the oven to 425°F/220°C (Gas Mark 7).
4. Return the dough to the work surface and stretch to an oblong. Fold in the dried fruit and continue kneading to distribute it evenly. Cut the dough into 16 equal sized pieces and form each into a roll. Place on the prepared baking sheets with space between to allow them to expand. Cover with a clean cloth and leave to rise and double in size.

5. Roll out pastry trimmings and cut into thin strips. Arrange them on the risen buns to make a cross and glaze with egg or milk.
6. Bake for 15 minutes. When the buns are cooked they will sound hollow when tapped on the base. Cool on a wire cooking tray.

**Optional Extras**

Half the buns can be decorated with oats by sprinkling two of the sections made by the pastry trimming cross with oats. This is done after the buns have been glazed so that the oats stick to the buns. Why not do half the batch in this way to make it more attractive?

The pastry trimmings can be removed from the buns after they have been baked to leave crosses in the buns. If doing this glaze the buns with either clear honey dissolved in a little boiling water or no-added-sugar jam – either hot or cold.

Whether you remove the pastry strips or not the buns always look glossier and more attractive if lightly brushed with the glaze after they have been baked or just before serving. Use an ordinary pastry brush to do this.

# EASTER MUFFINS

Makes 18 1 g fibre/70 calories

| | *Metric* | *Imperial* | *American* |
|---|---|---|---|
| *Wholemeal flour* | *100 g* | *4 oz* | *1 cup* |
| *Baking powder* | *2 tsp* | *2 tsp* | *2 tsp* |
| *Mixed spice* | *½ tsp* | *½ tsp* | *½ tsp* |
| *Ground cinnamon* | *½ tsp* | *½ tsp* | *½ tsp* |
| *Oat flour* | *50 g* | *2 oz* | *½ cup* |
| *Soft vegetable margarine* | *50 g* | *2 oz* | *½ cup* |
| *Currants* | *50 g* | *2 oz* | *½ cup* |
| *Sultanas or raisins* | *25 g* | *1 oz* | *2 tbsp* |
| *Clear honey* | *1 tbsp* | *1 tbsp* | *1 tbsp* |
| *Skimmed milk* | *150 ml* | *¼ pint* | *⅔ cup* |
| *Free-range egg* | *1* | *1* | *1* |

# Easter Muffins continued

1. Lightly oil a muffin pan or bun tray to take 18 buns. Set oven to 400°F/200°C (Gas Mark 6).
2. Sieve together the flour, spices and stir in the oat flour.
3. Rub the margarine into the flour until the mixture resembles breadcrumbs in consistency.
4. Stir in the dried fruit.
5. Beat together the honey, milk and egg and stir into the dry mixture. Then spoon into the prepared tin and bake for 15–20 minutes until just firm to the touch. Serve hot.

# EASTER BISCUITS

Makes 24 1 g fibre/75 calories per biscuit

| | *Metric* | *Imperial* | *American* |
|---|---|---|---|
| ***Soft vegetable margarine*** | ***100 g*** | ***4 oz*** | ***½ cup*** |
| ***Brown sugar*** | ***50 g*** | ***2 oz*** | ***¼ cup*** |
| ***Free-range egg, lightly beaten*** | ***1*** | ***1*** | ***1*** |
| ***Wholemeal flour*** | ***150 g*** | ***5 oz*** | ***1¼ cups*** |
| ***Oat flour*** | ***75 g*** | ***3 oz*** | ***¾ cup*** |
| ***Ground cinnamon*** | ***½ tsp*** | ***½ tsp*** | ***½ tsp*** |
| ***Ground mixed spice*** | ***½ tsp*** | ***½ tsp*** | ***½ tsp*** |
| ***Currants*** | ***50 g*** | ***2 oz*** | ***½ cup*** |
| ***Milk, brandy or rum*** | ***1 tbsp*** | ***1 tbsp*** | ***1 tbsp*** |

1. Lightly oil 2 baking sheets and set the oven to 350°F/180°C (Gas Mark 4).
2. Beat together the margarine and sugar until light and fluffy. Gradually add the egg.
3. Sieve the flour together with the spices and fold into the mixture together with the currants.
4. Add the milk, brandy or rum and lift the dough on to a lightly floured surface to roll out. To give the biscuits an attractive topping sprinkle some oats on to the surface and roll the biscuit dough out on top of the oats. Use a biscuit cutter to cut the shapes and lift them on to the prepared trays, oat side uppermost.
5. Bake for 15 minutes. Remove from oven and leave to cool and crispen on a wire cooling tray. Store in an airtight tin.

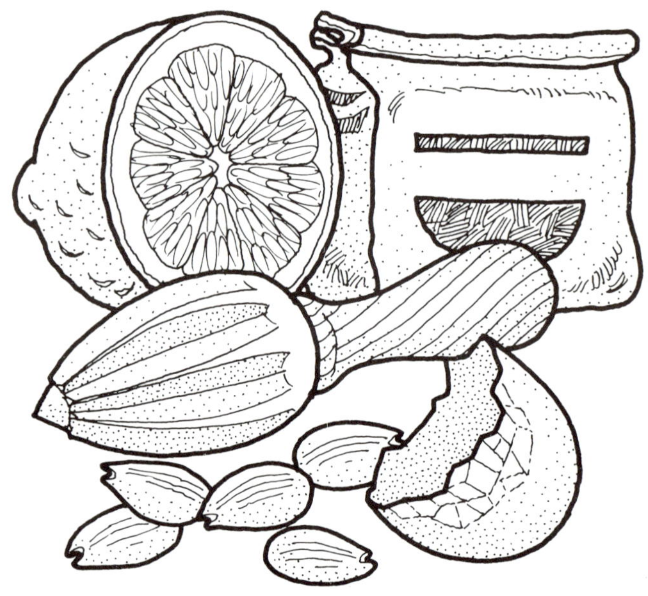

# Simnel Cake

Serves 12 3 g fibre/280 calories serving

| | *Metric* | *Imperial* | *American* |
|---|---|---|---|
| ***Soft vegetable margarine*** | ***175 g*** | ***6 oz*** | ***¾ cup*** |
| ***Raw cane sugar*** | ***100 g*** | ***4 oz*** | ***½ cup*** |
| ***Clear honey*** | ***2 tbsp*** | ***2 tbsp*** | ***2 tbsp*** |
| ***Free-range eggs, lightly beaten*** | ***3*** | ***3*** | ***3*** |
| ***Wholemeal flour*** | ***175 g*** | ***6 oz*** | ***1½ cups*** |
| ***Oat flour*** | ***50 g*** | ***2 oz*** | ***½ cup*** |
| ***Mixed spice*** | ***2 tsp*** | ***2 tsp*** | ***2 tsp*** |
| ***Nutmeg*** | ***½ tsp*** | ***½ tsp*** | ***½ tsp*** |
| ***Currants*** | ***50 g*** | ***2 oz*** | ***½ cup*** |
| ***Raisins*** | ***100 g*** | ***4 oz*** | ***1 cup*** |
| ***Sultanas*** | ***100 g*** | ***4 oz*** | ***1 cup*** |
| ***Raw cane sugar marzipan*** | ***225 g*** | ***8 oz*** | ***2 cups*** |
| ***Blanched almonds*** | ***225 g*** | ***8 oz*** | ***2 cups*** |

1. Lightly oil and line an 18 cm (7 inch) cake tin and heat oven to 325°F/170°C (Gas Mark 3).
2. Cream together the margarine, sugar and honey. Beat in the eggs, a little at a time. Sieve the flours with the spice and fold in. Stir in the fruit and place half the cake mixture in the prepared tin.
3. Roll out the marzipan to the circumference of the tin and place on top of the mixture in the tin.
4. Top with the remaining cake mixture and place the 11 almonds (to represent the apostles) around the edge of the cake.
5. Bake for 1½–1¾ hours until inserted skewer comes out clear. Top with double layer of greaseproof paper towards the end of cooking to prevent the cake over-browning.

# INDEX

alcohol 18
Apple Crumble 68
Apple Streudel Crêpes 71–72
Apple Streudel Ice-Cream 70–71

Banana Bread 91–92
Banana Crêpe 81
beans, *see* pulses
Beef Burgers 61
bile acids 10
Black Cherry Fool 78–79
Blackberry and Apple Crumble 69
Blinis 108
blood pressure, high, 8, 10
Braised Meatballs and Vegetables 57
bran 7, 8, 12
  oat 9, 13, 15–16
Breakfast Scones 36
breakfasts 18–19, 27–40
butter 8

carbohydrates 11
Chantilly Yoghurt Cream 80
cheese 8, 11
Cheese Crackers 101
Cheese Flapjacks 100
Cheese and Onion Pancakes 105–106
Cheese Scones 101–103
Chelsea Buns 88–89
Chicken Chervil Croquettes 42–43
cholesterol 9, 10, 13
Christening Biscuits 109
Christmas Cake 117–118
Christmas Pudding 115–116
Cinnamon Crumble Cake 94–95
Coconut Muesli 27
Coffee Crunch 110–111
constipation 7
Corn Muffins 33–34
Crêpe Suzettes 69–70
Crunchy Bars 99–100
Crunchy Carob Biscuits 112–113
Currant Cheesecake 73

Date Slice 97–98
Date and Walnut Loaf 85
diabetes 9, 13
Dr. Anderson's Oat Muffins 28
Dutch Apple Cake 75–76

Easter Biscuits 123–124
Easter Muffins 121–122
Eastern Fishcakes 43–44
Easy Croissants 39–40
eggs 11, 19
exercise 18, 23–24

Family Favourite Granola 37–38
fats 8, 10, 11, 17
fibre 6–14, 17, 25–26
Fig Muffins 30
fish 11, 19
fruit 18, 19, 23, 26

Fruit Cake 89–90
Fruit and Nut Muesli 30
Fruit Salad 76–77
Fruit Tart 66
Fruity Porridge 37

gallstones 8, 10
Ginger Biscuits 114
Gingerbread 86–87
granola cereal 18

Hazelnut and Carob Squares 113
health farm 21
health weekend 21–24
heart disease 8, 9, 13, 14
Herb Scones 107–108
Homemade Pasta 60
Honey Flapjacks 105
Honey Oaties 111
Hot Cross Buns 120–121
Hot Fruit Compôte 77
Hot Waffles 32–33

jumbo oats 16

Leek Pie 52–53
Lemon Curd Tart 65–66
Lime Cheesecake 74–75

Madeleines 87–88
Malt Loaf 103–104
Marzipan Franzipan 79–80
meat 7, 8, 11, 12, 13
milk 8, 11, 19
Mince Pies 118–119
minerals 13, 17, 18
muesli 18, 20, 22, 27, 30
Mushroom and Cashew Lasagne 58–59
Mushroom Fish Pie 55–56

No-Sugar Rock Buns 93
nuts 19, 22, 26

oat bran 9, 13, 15–16
oat bread, 14, 20
Oat Bread 93–94
oat germ 15–16
oat muffins 13, 14, 18, 28, 32
oat pancakes 18, 20, 48
Oat Pancakes 48
oat pastry 20, 47
Oat Pastry 47
oat scones 14, 18, 36
oat toppings 19, 20
Oat Vegetable Pie 63
oat waffles 32–33
Oatcake Fingers 106
oatcakes 19, 20
Oatcrunch Coating 42–43
oatflakes 16
oatmeal 15, 16
  stabilisation of 16
  storage of 16
Oatmeal Biscuits 110
oatmeal breads 19, 93–94
oaty crunchy bars 19, 99–100
Onion Tart 48–49
Orange and Apricot Cheesecake 72–73

Pancake Filling 46
Parsley Scone Dumplings 60–61
Parsnip Croquettes 56
Piped Oaties 112
Piping Chantilly Cream 80–81
Pizza Scone 44–45
Plain Porridge 35
porridge 35
porridge oats, *see* oatmeal
protein 11, 17
pulses 7, 8, 11, 12, 17, 22, 23, 26

refined foods 12
relaxation 24

Rhubarb Crumble 67–68
Rich Croissants 29
rolled oats 16

salads 19, 20, 21–22
salt 17
Savoury Snaps 96
Scotch Eggs 96–97
Sesame Flapjacks 104
Shortbread 83
Simnel Cake 125
slimming 8, 23
smoking 18
Soda Bread 85–86
Soft Oat Muffins 32
Soluble Fibre Table 25–26
Spiced Crispy Chicken 59
Spinach Pancakes 41
Spinach Roulade 50–51
Spinach Soufflé 49
Spinach and Yoghurt Soup 51–60
stabilisation, oatmeal, 16
Strawberry Shortcakes 82
sugar 12, 17, 18
Sultana Scones 92
Swede and Carrot Gratin 61–62
Sweet and Sour Sauce 46

Toasted Teacakes 84
Traditional Moussaka 64

Unprocessed foods 11, 12, 17

Vegetable Burgers 62
vitamins 13, 17, 18

Walnut Moussaka 53–54
Wheatgerm Muffins 31

Yoghurt Treat 106